Ahtna Travel Narratives

A Demonstration of Shared Geographic Knowledge
among Alaska Athabascans

ANLC

ALASKA
NATIVE
LANGUAGE
CENTER

Alaska Native Language Center
Box 757680
University of Alaska
Fairbanks, AK 99775-7680
http://anlc.uaf.edu

First printing   2010   1,250 copies

Elmer E. Rasmuson Library Cataloging in Publication Data:
Ahtna travel narratives : A demonstration of shared geographic knowledge among Alaska
Athabascans / told by Jim McKinley [et al.] ; transcribed and edited by James Kari. Fairbanks,
Alaska : Alaska Native Language Center, University of Alaska Fairbanks, c2010.
  p. cm.
Includes bibliographical references.
ISBN 978-1-55500-105-6
1. Ahtena Indians-Travel—Alaska. English language—Dictionaries—Yuit. I. McKinley, Jim,
1898-1989. II. Kari, James M.
E99.A28 A48 2010

Cover photo:	View of the Copper River as ice broke up, from below the mouth of the Gakona River [1-54, Ggax Kucae'e] with Mount Sanford [4-39, Hwniindi K'ełt'aeni] and Mount Drum [Hwdaandi K'ełt'aeni] in the distance. Photo by Suzanne McCarthy.
Cover map:	Henry T. Allen, Map no. 1, 1887. Details on pages xiv and 81.
Back cover photo:	"Bear John." Details on page xv.
Cover design:	Dixon Jones, UAF Rasmuson Library Graphics.

Ahtna Travel Narratives

A Demonstration of Shared Geographic Knowledge among Alaska Athabascans

Told by
Jim McKinley
Frank Stickwan
Jake Tansy
Katie John
Adam Sanford

Transcribed and edited by
James Kari

Alaska Native Language Center
University of Alaska Fairbanks
2010

Contents

Symbols and Abbreviations

¶   paragraph break
/   free English translation of Ahtna sentence
[...]   word added that was not recorded or clarification in translation
1:31   time count, indicating minutes and seconds in the audio file
f.s.   false start (incomplete word, replaced by an altered word)
pl.   plural
ques.   question word or question particle
sg.   singular
sth.   something
p.c.   personal communication
Ck   Creek
L   Lake
Mt   Mountain
R   River

Place name citation style
Ts'es K'et (with capitalized words)
'on the rock' (translation in single qoutes)
[1-14, bluff on W bank above Tonsina R]
1-14 (place name reference number) the 14th place mentioned in Chapter 1

Ce's Di'aedzi Cae'e
/'? spear stepping mouth'   '?' = meaning is speculative

Sdzedi Na'
'sdzedi stream'   'sdzedi' = meaning is uncertain

Acknowledgments

Many Ahtna speakers have contributed to the general knowledge of Ahtna geography, and a list of the known contributors is in *Ahtna Place Names Lists* (Kari 2008:vi). It is with great appreciation that I present, by dialect area, the main contributors I have worked with:

Lower Ahtna: Jim McKinley, Andy Brown, Frank Billum, John Billum, Robert Marshall, and Walter Charley.

Central Ahtna: Martha Jackson, Fred Ewan, Ben Neeley, Buster Gene, Jim Tyone, Frank Stickwan, Oscar Ewan, Andy Tyone, Markle Pete, and Tenas Jack.

Upper Ahtna: Katie John, Fred John Sr., Kate Sanford, Jack John Justin, Bell Joe, and Adam Sanford.

Western Ahtna: Henry Peters, Jake Tansy, Dick Secondchief, and Johnny Shaginoff.

For the work on this book I thank especially Markle Pete, Fred Ewan, and Katie John. I also appreciate the help and support of many people at Ahtna Inc., including Kathryn Martin, Ezekial Beye, Joe Bovee, Kenny Johns, and Eileen Ewan. I also thank friends and colleagues Bill Simeone, Siri Tuttle, Gary Holton, Andrea Berez, Rick Thoman, Suzanne McCarthy, Molly Galbreath, Ben Potter, Geoff Bleakley, and Adeline Kari for help with various aspects of this book. Leon Unruh at the Alaska Native Language Center has done a fine job with the layout and editing of the book.

I acknowledge with thanks several funding sources that have contributed to this book:

During 1998–2004 funding for Ahtna fisheries research through the Alaska Department of Fish and Game and the U.S. Fish and Wildlife Service contributed to Ahtna geographic research. In 2005 a contract through the Bureau of Land Management for the Eastern Alaska Management Plan also contributed to geographic research. During 2007–2009 research for and layout and printing of the book was sponsored by a grant from the National Science Foundation #0553831 "Ahtna Texts" (Siri Tuttle, principal investigator).

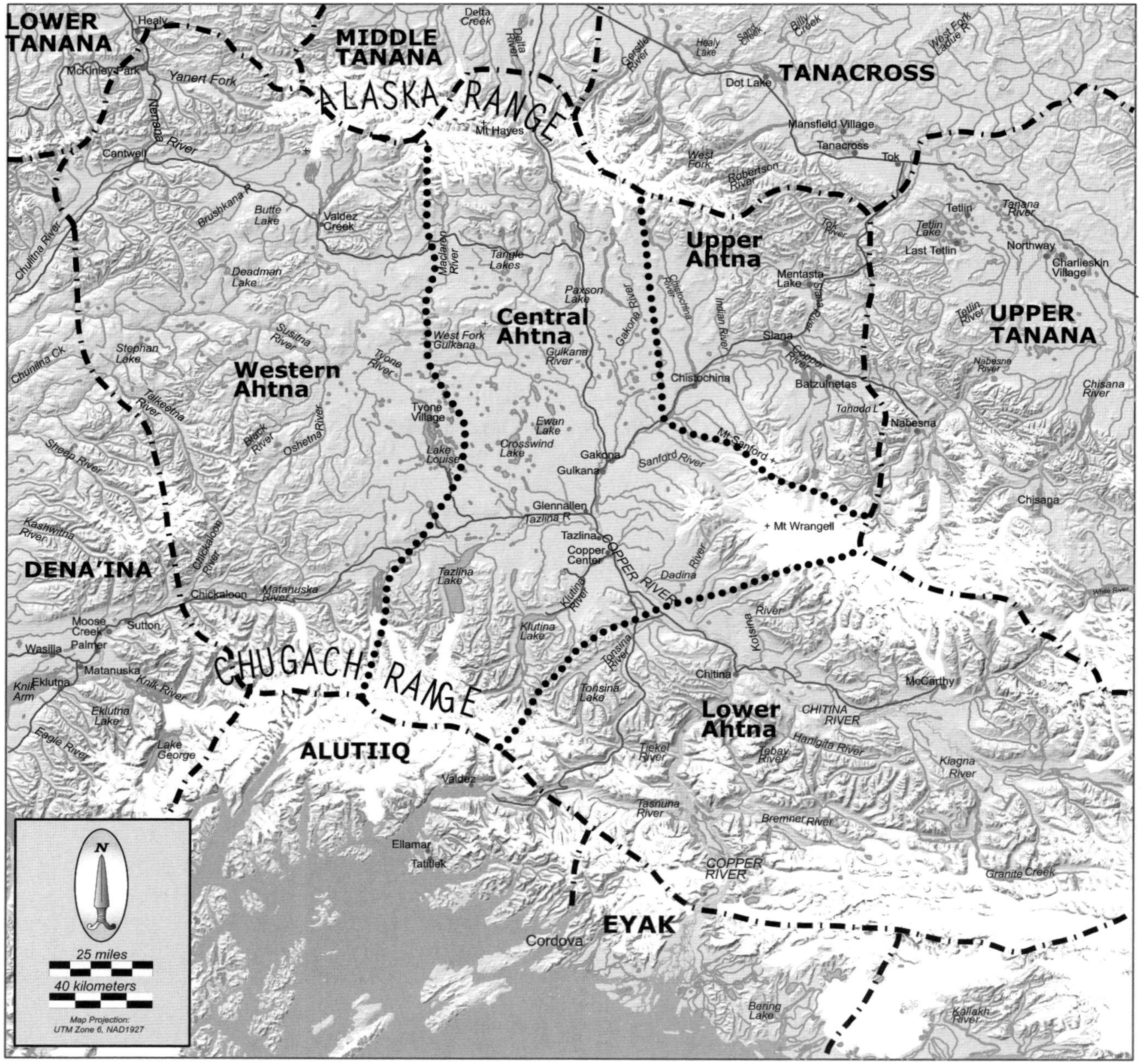

Ahtna Language Area

language boundaries

dialect boundaries

• Chistochina
settlements

Ewan Lake Mt Sanford
place names

modern roads

Base map by Matt Ganley, Map-Alaska
Text and Ahtna names by James Kari

Figure 0-1. The Ahtna language area showing the language and dialect boundaries.

Introduction

Figure 0-2. One of the best places to view a large part of the Ahtna language area is from Hogan Hill on the Richardson Highway. This 2,647-foot hill, named in Ahtna **K'ey Tsaay Gha** 'by the small birch', was an island when Glacial Lake Ahtna filled most of Copper River Basin prior to breeching more than 10,500 years ago. The Wrangell Mountains are called **K'eɬt'aeni** [1-25], a "high word" in Ahtna that is difficult to translate into English. Mount Sanford (to the left) is **Hwniindi K'eɬt'aeni** [4-39], and Mount Drum (center) is **Hwdaandi K'eɬt'aeni.** The names mean 'upriver K'eɬt'aeni' and 'downriver K'eɬt'aeni', respectively.

Photo courtesy of University of Alaska Fairbanks Geophysical Institute.

Yenidan'a ts'en koht'aene tene kulaen de.
/The Ahtna people's trails have existed since the ancient times.

—Katie John

Nts'e ye hwdi'aandze' hw'eł ts'etnes.
We know the place there by how it is named.

—Fred Ewan

The five narratives in this book are a set of walking tours of traditional Ahtna lands by five experts: Jim McKinley, Frank Stickwan, Jake Tansy, Katie John, and Adam Sanford. This book and a second book in preparation are a selection of Ahtna language narratives that represent the favorite genres (or subject areas) of the Ahtna oral tradition.

Today's Ahtna elders have had many opportunities to hear traditional storytelling from the prominent Ahtna storytellers and raconteurs of 30 to 50 years ago. The favorite genres of the Ahtna oral tradition can be summarized:

  (a) yenida'a stories, the stories of mythic times when men and animals spoke
  (b) major events in Ahtna history (ancient to recent): clan origins, stories about well-known

       people, wars and altercations, early contacts with white people

(c) discussions about cultural practices and lifeways, such as copper, houses, boats, about fishing or fish preparation; discussions about customs and beliefs

(d) songs and stories about songs

(e) stories about travel and place names

A second book, *A Selection of Ahtna Yenida'a (Legendary) and Cultural Stories* (Tuttle and Kari, to appear), is a sample of genres a–d. **Yenida'a** means 'long ago' in Ahtna, and **yenida'a** stories are set in a time of madcap events in which animals and humans talk in Ahtna. It is noticeable that the Ahtna **yenida'a** myths always lack place names or any local geographic references. The two published collections of **yenida'a** stories by John Billum (1979) and Jake Tansy (1982) can be considered pure fiction, and they contain no place names. On the other hand, the presence of place names and actual geographic locations is a defining trait of nonfiction Ahtna narratives. The book of Upper Ahtna narratives (Kari 1986) contains nonfiction stories that take place at named locations in the Ahtna landscape; these stories include two incidents in which Russians were killed, and the fantastic stories about Cet'aeni, 'the tailed ones' or the "monkey people," as well as the giant fish stories that take place at several large lakes.

Ahtna is a member of the Athabascan language family, the largest indigenous language family in area in North America. The Ahtna language area is very large, covering about 35,000 square miles, and is centered mainly on the Copper River, a 250-mile-long stream that heads on the north slopes of the Wrangell Mountains and flows into the Gulf of Alaska. As shown in Figure 0-1, the Ahtna language includes all of the Copper River drainage above Childs Glacier; in the northeast: most of the Tok River drainage; in the north: the upper Delta River to the Black Rapids area; in the northwest, the upper Susitna River above Devil's Canyon and the upper Nenana River above Healy River; and in the west, the upper Matanuska River above Chickaloon River. *Ahtna Place Names Lists* appeared in 2008 in a revised second edition with 2,208 names in 20 sections organized by drainage. The 36-page introduction has discussion of the sources and the key features of Ahtna place names.

In the early history of exploration of Interior Alaska, the travel skills and geographic expertise of Athabascan leaders had an enormous (and largely unappreciated) impact on trail reconnaissance, cartography, and place naming. The skills and the breadth of the Athabascan geographic expertise in eastern Alaska is readily seen in the best sources of this period (such as Allen 1887; Rohn 1899, 1900a, 1900b; and Powell 1910). By 1900 about 140 Ahtna place names had been recorded, and many appear on the earliest maps of the Copper River area. Several Ahtna place names have been reported consistently for more than 200 years.

The journey of U.S. Army Lieutenant Henry T. Allen up the Copper River in the spring of 1885 was a pivotal time in Ahtna and Alaska history. Allen's report, published in 1887, is remarkable for its descriptive detail. (The original book is rare, but in recent years it has been available as an electronic file.) The Ahtna population was never very large before historic contact. In 1885 Allen questioned Ahtna leaders he encountered, and he estimated the Ahtna population to be 376. The first U.S. census in Copper River in 1910 counted only 297 Ahtna people (Bill Simeone, p.c.). The fact that such a comprehensive network of Ahtna place names extends throughout and beyond the 35,000-square-mile language area is a significant demonstration of the prodigious travel prowess and geographic expertise of the small Ahtna or Athabascan population that has occupied the Copper River Basin for a very long time (at least 6,000 years but likely for more than 10,000 years).

This shared geographic knowledge as attested in networks of Ahtna place names has many regular and formal features. In Kari 2008 these points are exemplified in more detail. For Ahtna we can marvel at the strict purity, orderliness, symmetry, redundancy and functionality of the geographic names. We can make many generalizations based on the Ahtna place names corpus. Numerous features of the place names facilitate memorization. The place names can be summarized in terms of information content, structural patterns, distribution, use in overland navigation, and many other features. The place names follow several structural patterns and use both simplex nouns and complex verbs. The specific names function as signs, with clear analyzable meanings. There are virtually no non-Ahtna or non-Athabascan place names. Many of the specific names generate into clusters or sets of three or four names that apply to a drainage, a hill, or a mountain. These bi- and trinomial place names employ regular generic terms that classify the landscape. The order of streams is the key organizing principle, and there is a low density of names.

Anyone who has spent time with Ahtna speakers knows that stories about Ahtna place names and travel on the land are subjects of endless fascination. I use the term *elite travel narrative* to refer to connected text in the language that describes with precision sequences of places or travel in a region that the speaker knows well. The five narratives presented here are a fraction of the various Ahtna travel narratives recorded with Ahtna elders over the past 30 years. In these walking tours of traditional Ahtna lands, the places that are named and the routes that are summarized can be translated, traced, and mapped fairly closely.

Travel narratives are a neglected and rare narrative genre in Amerindian literatures and in indigenous literatures worldwide. This is due to a combination of factors. The few people who can describe travel and landscape really well must have special athletic abilities and travel skills. The translation of travel narratives is labor intensive. For a variety of reasons, there is real value and importance to a careful presentation of narratives about Ahtna travel (see Kari 2004). These travel narratives show the shared Ahtna geographic knowledge "in action" as no other texts or interviews can. Today Ahtna lands are not being accessed and used as they were in 1900 or even in 1950. These travel narratives offer a wealth of information about Ahtna geographic names, specific streams and landforms, foot travel, trails, and resource use.

Texts about places and travel are important complements to place names research. Place names need to be consolidated in river drainage files so that names can be added or refinements in locations can be made. The editing of travel narratives requires the mapping of locations and routes as well as consistency in the translation of the meanings of place names and of the elaborate riverine directionals. In the course of preparing these five narratives, quite a large number of refinements were made beyond the names presented in the 2008 second edition of *Ahtna Place Names Lists*. The daunting editorial challenges presented by elite travel narratives can be seen in the Katie John and Adam Sanford narratives, which were published previously in 1986. The current versions have many improvements in the transcription of the Ahtna sentences, transla-

Figure 0-3. Jacksina Creek viewed from the west bank.

Lower Jacksina Creek is called **Dimee Jiidi Ndiig** 'rotten sheep creek' in the Nabesna dialect of Upper Tanana. In the distance is the Nabesna River—its Ahtna name is **Nabaes Na'** [5-62], and in Upper Tanana it is **Naabia Niign.** The upper Nabesna River valley is the home area of the Nabesna people, who speak a dialect of Upper Tanana. There are numerous bilingual place names for features in this area.

Some of the best descriptions of the geology and trail systems in the Copper River from the 1890s appear in articles by topographical engineer Oscar Rohn:

"The Nabesna divide is cut by three easy passes, suitable alike for horse trail and railroad. The Nabesna-Tanana divide is crossed by a pass somewhat more difficult, but the Tanana-White divide is merely a range of hills. The area is, therefore, very accessible, and affords an easy route from the

valley of the Copper to the valley of the Yukon. . . . Between these and the Mentasta range, and north of the Nebesna divide, is the head of the Copper River Valley, a rather flat area studded with innumerable lakes and bogs. Several lakes attaining considerable size are named by the natives Tanada, Zachnada, Totrachara, and Suslota. Streams from these lakes drain into the Copper River" (Rohn 1900c:792).

The Ahtna place names obtained by Rohn are **Tanaade Menn'** [Tanada Lake 4-21], **Dzah Nii Menn'** [Copper Lake, 5-60], **Tadiniłts'aegge Menn'** [Jack Lake], and **Sasluugge' Menn'** [Suslota Lake]

Photo taken on July 29, 1938, by T. W. Ranta, U.S. Geological Survey. Courtesy of Bill Simeone.

tions, the locations of many specific place names, and overall editorial consistency.

In these narratives of travel there is an orchestration of several types of linguistic systems. The place names are a highly specialized lexicon that is used in combination with several other systems. There are nouns and verbs that classify landforms, vegetation, and hydrology. Recurrent verb themes characterize the trajectory and appearance of ridges, mountains, and trails, or dynamic hydrological conditions. The subtleties of the person walking through the landscape are encoded in the grammar of the motion verbs. There is also a demonstrative system (**gaa** 'here', **yet** 'there', **duugh** 'there in the distance'). Uniquely, for Ahtna and other Northern Athabascan languages, there is the *riverine directional system,* which contains nine basic roots and which is constantly applied as a frame of reference. For most of Ahtna the absolute frame of reference is the Copper River. (Among the Valdez Creek–Cantwell Ahtna, the Susitna River is the major drainage.) The riverine directionals triangulate with the Ahtna geographic names to track routes in relation to the major stream.

The editing of these narratives makes some compromises on details but strives for consistency, in structure, meanings, and repetition in the use of the Ahtna places names and the riverine directionals. Chapter 6 contains a summary of the Ahtna riverine directional system and the use of place names and riverine directionals in the five chapters.

Numerous illustrations of Ahtna geographic expertise and travel prowess are noted throughout the book in footnotes or in captions for photographs, historic maps, sketch maps by various Ahtna speakers, and pages from some of my field notebooks. Also, there is discussion of some of the methods I employ to research and to annotate Ahtna geography.

The conventions in this book for translation of place names, landscape terms, and riverine directionals give a sense of the functionality of the shared orally transmitted navigational system. Place names are numbered in the order that they are mentioned by a *place name reference number.* The repeated mention of place names—by the same speaker or by different speakers—is noted throughout. When a previously mentioned place name is repeated or is used in a figure, this number is noted.

The details and facts in these narratives certainly must place this collection among the most detailed examples of foot travel ever published in the language of a hunter-gatherer culture. In the five chapters, in just over 102 minutes of speech nearly 1,200 miles of trails and routes are described; 313 distinct place names (Ahtna names for different places) are mentioned 698 times by the same narrator. The complicated riverine directional terms are used 573 times. Reiterated place names (defined as the same place name mentioned by a different speakers) occur 41 times in the five narratives.

In two segments of 45 minutes recorded on the same day in 1981 for about 280 miles of trail routes, Jim McKinley mentions 118 different place names 375 times and uses 231 directional terms. In one segment of 3 minutes 8 seconds, Jake Tansy gives a detailed summary of alternate routes of about 125 miles between the Nenana River and Yanert Fork and back to the Upper Susitna River. In this segment, he mentions 22 place names 36 times, and he frames these with 26 directional terms. Tansy's narratives are masterful examples of topographic description and intimate familiarity with the landscape. Chapter 6 presents more facts

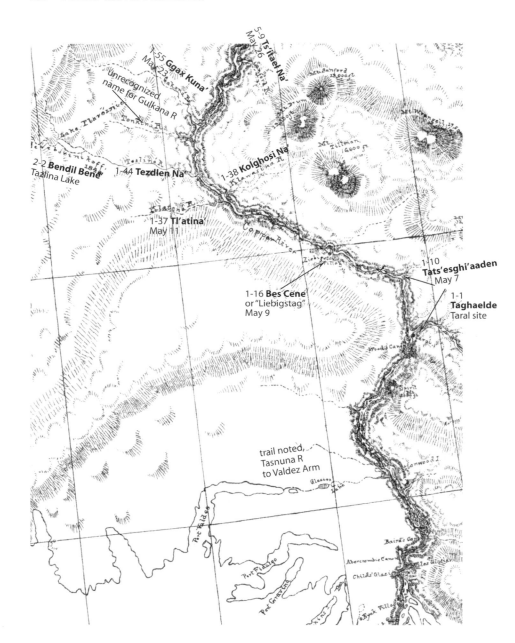

Figure 0-4. Close-up of the central Copper River from Allen 1887, Map no. 1 (also on this book's cover and in Figure 4-3). Noted on the map are nine Ahtna place names with reference numbers as well as the dates in May 1885 that the Allen party passed these places.

Between March and August 1885, Henry T. Allen's U.S. Army party ascended the Copper River, rafted down the Tanana River, walked over to the Upper Koyukuk River, and then rafted down to St. Michael at the mouth of the Yukon River. The five maps made by Lieutenant Allen, Sergeant Cady Robertson, and Private Frederick W. Fickett are considered the most accurate maps of Alaska prior to the professional mapping by the U.S. Geological Survey in the late 1890s. The fine detail on Copper River geography and on Ahtna sites, trails, and place names pays tribute to the "shared knowledge" and geographic expertise of the Ahtna leadership who facilitated the travel of the Allen expedition. About 50 Ahtna place names appear in the Allen report and maps.

Allen (1887:117) wrote about the maps: "I think the great care taken to secure a correct description of the rivers will be of great value. . . . Each of the maps is constructed on a polyconic projection from tables published by the Bureau of Navigation, and . . . on a scale if 1 inch to 4 miles."

about the use of place names and directionals.

The stories are presented in a two-line format. The first line is an Ahtna sentence or phrase. The second line, following /, is a translation of the phrase in regular English phrasing. Audio files for the narratives are on an accompanying CD in six files. Time marks are noted throughout the narratives so that the audio file can be followed or paused and replayed. When you play this file on a computer or CD player, you can listen to and study these lines in several ways. You can play the entire track, or you can pause the file after a few seconds and then read or practice what you have heard. In programs such as Audacity, Sound Forge, and Cool Edit, there are ways to replay short sections of sound, such as one word or phrase or a sentence. Also, some avid learners of Ahtna and other Alaska Native languages have been playing these audio files on their iPods. We have made note of "false starts," which are abbreviated *f.s.* These are incomplete words started by the speaker and then altered or revised in the next word of the sentence. Some readers may find these false starts to be of interest.

Ahtna geographic names are so informative and learnable that they facilitate the understanding and recognition of the landscape. To be sure, all of the Ahtna names in this book can be used and committed to memory.

Figure 0-5. Man identified as "Bear John" on the Cheshnina River (Tsesnen' Na'), **which the Allen map labeled as Liebistag River [0-4] in 1905–1910.**

Note the well-worn trail. Robert Marshall thinks this man looks like John Billum Sr., son of the famous Doc Billum. The Billums lived in the area opposite the mouth of the Tonsina River and were very active in commerce in the first decades of the century. Robert thinks that "Bear John" was a nickname given to John Billum by white people. Robert Marshall's stepdad, Jack Marshall, had the nickname "Bear Jack."

1

'Atna' K'et Kayax 'eł Tl'atina' Ngge'

Ahtna Villages on the Copper River and in the Klutina River Drainage

Jim McKinley

Figure 1-1. Jim McKinley in Fairbanks in January 1981, when these narratives were recorded.

Photo by James Kari.

Jim McKinley was born in Copper Center in 1898 and died in 1989. Jim held leadership roles throughout his life as a clan leader, storyteller, singer, interpreter, and ordained minister. Jim, and before him his father, McKinley George, maintained cabins, caches, and other structures on Klutina Lake at the mouth of Mahlo River. The Ahtna elders have a custom of caucusing to discuss regional geography, history, genealogy, and songs, and when I first came to Ahtna country in the 1970s and 1980s Jim McKinley was one of the revered sources on cultural history for the Ahtna region. He knew details about Ahtna history, travel, and sites all along the Copper River and also throughout and beyond the Ahtna language area. Jim was most intimately knowledgeable about the Klutina River drainage because this had been his family's main area for hunting and trapping. In the 1980s he was named the first traditional chief of the Ahtna Region.

I worked with Jim McKinley on about 20 occasions between April 1974 and October 1989, and we recorded about 75 pages of notes and about three hours of audio recordings. On January 12, 1981, Jim and I tape-recorded about 49 minutes of geographic narratives that were retrospective of our previous work.

These two "walking tours" of Ahtna territory contrast nicely. The first narrative is like a well-prepared academic lecture. It is a beautifully paced 27-minute survey of the major Ahtna villages

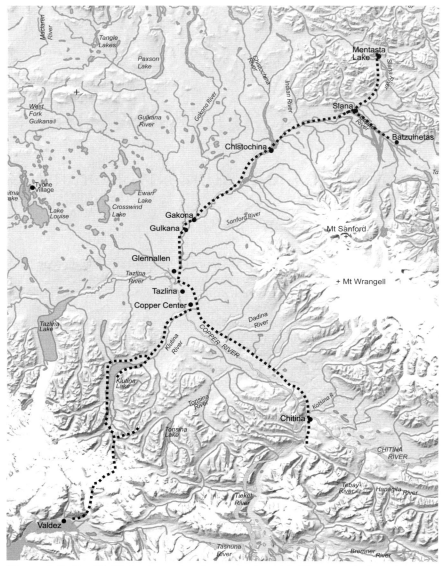

located along the Copper River. Jim highlights and explains 71 of the best-known Ahtna village sites, in an upstream succession, from below Taral to Mentasta, a distance of at least 160 miles. The narrative is a bird's-eye view of the settlements as of approximately 1875. At one point Jim notes in English, "That's where people living, that's all I talk about." In other words, Jim refrained from mentioning side streams and landmark features not pertinent to the sequence of Ahtna village sites.

In Part 2 Jim presents a sequence of names in his family's primary land-use area, the Klutina River and Klutina Lake area as well as the trail to Valdez over the Klutina and Valdez glaciers. Part 2 is a walking tour, with names given along the trails in ascending order with a few deviations.

Figure 1-2. Jim McKinley's routes for parts 1 and 2. The route in Part 1 up the Copper River is at least 160 miles long. The route up the Klutina River toward Valdez in Part 2 is about 110 miles.

Part 1
Ahtna Villages on the Copper River

Starting point: above Wood Canyon on Copper River
Ending point: Mentasta
Rec. Jan 12, 1981, AT 23 or AT 5440; total: 27:34. Sound file chp1-1-jimmckinley.wav.

Saen hdaghalts'e'de tah, you know.
/Where they stayed during the summer, you know.

'Udaat Taghaelde kudaa'
/Downstream, downstream from Taral [1-1 Taral]

Tats'abaelghi'aax Ghaxen dae' konii, you know.
/was (the chief named) 'the person of spruce extends into the water', thus he was called [1-2, site on east bank at 119 mi.].

Yidi gha' hwdi'aan łdu' 'udo'ohniide Tats'abaelghi'aax Ghaxen?
/Why do you think he was called 'the person of spruce extends into the water'?
:30
You know that ts'abaeli taghi'aayi yii gha' su.
/You know there is a spruce sticking into the water, due to that.

Ts'abaeli taghi'aayi, only place, you know.
/A spruce extends into the water, that's the only place, you know.

'Udaa'a yedu' last you know, 'udaa'a dghilaayi dighiłeni xunt'aeyi.
/Toward downstream that's the last place; downstream (from there) there are mountains with swift current flowing.

Yii gha' su Tats'abaelghi'aax Ghaxen dae' konii. Tats'abaelghi'aax.
/This is why he (the chief here) was called 'the person of spruce extends into the water'.[1] 'A spruce extends into the water'.

That's spruce, yii su 'adetnii, spruce only place ts'abael taghi'aa de as far as down there.
/It is said that is the spruce, the only place the spruce extends into the water as far as that place.

Duu yihwts'en kaniit Tsenłt'e' Cae'e, Tsenłt'e', Tsenłt'e' Cae'e.
/From there the next place upstream is 'Tsenłt'e' mouth' [1-3, mouth of Eskilida Ck].

---

[1] This is a chief's title, one of four that Jim McKinley mentions in this narrative. As many as 16 titled chieftainships—where the chief of a major village was named for the places—have been documented (Kari 1986:15, Kari and Tuttle 2005:22–25). Nine of these titles were on the lower Copper River (below the Klutina River), which was the center of the majority of the aborignal Ahtna population. See also Notes on Ahtna Personal Names on pages 55–57.

Yełdu' sii cu u'e 'sdadestniic you know. Yidi c'a 'adetnii, something yii gha hwdi'aande dae' udetnii something, u'eł 'sdadestniic.
/That (name) I cannot understand what that means. There is some reason it is called that way, but I don't understand what it means.
1:07
Yet kanii yełdu' Ts'es K'et dae' konii, yełdu' uk'et ciisi t'al'iix yii gha' su, ciisi uk'e t'al'iix you know.
/Then the next place upstream then is called 'on the rock', because on that place there would be dipnetting, dipnetting would take place upon it [1-4, on west bank at mile 127].

Yii gha' su Ts'es K'et hwdi'aan, Ts'es K'et dae' hwdi'aan.
/This is why it is called 'on the rock', thus it is called, 'on the rock'.

Gha yełdu' kanii yełdu' Tsenłt'e' Cae'e dae' konii; yet yegha'aa Tsenłt'e' Cae'e.
/Now then the next place upstream is called 'mouth of Tsenłt'e'' [1-3, mouth of Eskilida Ck] and then out beyond it is 'mouth of Tsenłt'e''.[2]

Yełdu' xona yidi c'a k'adii yet su 'adetnii, yii cae'e dae' konii, dae' koniiyi.
/And then whatever that now is called (Eskilida Creek), that stream mouth is called that way, thus it is called.
1:45
Du' yet ka'aa yak'a xona Hwt'aa Cae'e dae' konii, Hwt'aa Cae'e
/Then there the next place beyond there is called 'enclosed mouth', 'enclosed mouth' [1-5, mouth of Fox Ck].

Yet su Hwt'aa Cae'e Denen ghida' dae' koniide.
/There is the place where the one called 'person of enclosed mouth' [second chief title mentioned] stayed.

Yii gha' su Hwt'aa Cae'e dae' hwdi'aan. You know what it means, Hwt'aa Cae'e?
/This is why it is called 'enclosed mouth'. Do you know what that means, 'enclosed mouth'?

 Hwt'aa Cae'e dzeł dzeł ghatgge dghilaay ts'idiniłen. Yii yits'ediniłen you know.
/'Enclosed mouth' flows out from between the mountains. It flows on out from within there.

'Uniide Chitina town kulaende xu cu dan'edze' dan'edze' bi'niłededlen xunt'ae
/Upstream of where Chitina town is, from the upstream, from the upstream, there these streams join.
2:18
One, one that little creek, one ts'ełk'ey łdu' dadaa'adze''Atna' yinaadlen xunt'ae.
/One is that little creek, one flows to the downstream into 'Atna' [1-6, the Copper R].

Ghadii cu yii łdu' dan'edze' 'uniit Nahwt'en Cae'e yihwk'e dighiłeni xunt'ae, dayggedze' ghayet dayggedze'.
/The other one flows from the upstream, to upstream at 'things reoccur mouth' [1-7, mouth of Fivemile Ck], from down below, there at that place down below.

---

[2] Jim inadvertently twice mentioned the site at the mouth of Eskalida Creek, which is actually below **Ts'es K'et.**

'Atna' yii another deyii, that's two, that's one creek creek niłts'idaadleni xunt'ae, you know.
/Copper River has another canyon, that's one drainage that divides apart, you know.
2:34
You see yii gha' su cu Hwt'aa Cae'e dae' hwdi'aan. Hwt'aa Cae'e see. Hwt'aatna' dae' koniide see,
/You see, that is the reason it is called 'enclosed mouth' [1-5, mouth of Fox Ck], 'enclosed mouth', the place (the creek) is called 'enclosed stream' [1-8 Fox Ck].

Tsabaey cu yii c'ilaeni, tsabaey cu łuk'ae yii talax, dae' nt'ae.
/There are also trout in there, also salmon run in there as well, that is how it is.
2:48
Dae' dghit'ae', you know, yii gha' su k'adii k'adii cu kukaet whiteman ik'ediłt'ae.
/That is how it was, the reason (how it was named), you know, and now the white man has bought the place and he uses it.

Tsabaey uyii talyaesi t'el'aeni, see.
/Fish are taken there, see.

Yii gha' su Nahwt'en Cae'e.
/The reason for 'things reoccur mouth' [1-7, mouth of Fivemile Ck].

Du' kanii Nahwt'en Cae'e yełdu' dae' su 'uniidze snaghideł de yet hwk'e yaen' yax deke' kae łuhghideł.
/The next place upstream, 'things reoccur mouth' is where they used to turn back around to come back upstream when they used to go around just by foot.

Dadaa'dze' łdu' cu you know it's hard, you know, all dghilaay tah, hwnene tah 'eł kulaen.
/(Coming) from downstream of there, you know, it is hard (walking) you know, among all the mountains and slopes.
3:15
Yii gha' su Nahwt'en Cae'e dae' konii de 'Atnahwt'aene yehwk'e ninaghedełi yii gha' su konii.
/Why it is called 'things reoccur mouth' is that Ahtna people used to regularly stop there, that is why it is called so.

Dae' Nahwt'en Cae'e, Nahwt'en Cae'e dae' koniiyi.
/Thus 'things reoccur mouth', 'things reoccur mouth' it is called.

Yii gha' hwdi'aani sunt'ae.
/That is why it is so named.

Yihwts'en kanaa 'uniit Tay'laxi Na'. Tay'laxi Na' deyii łdu' łuk'ae 'ungge una' talax, łuk'ae una' talaxi.
/From there the next place upstream is 'fish run stream'. 'Fish run stream' has a gorge and the red salmon run upland there in that stream, salmon run in that stream.

Yii gha' su łuk'ae Una' Tay'laxi Na' [or Tay'sdlaexi Na'] hwdi'aan.
/That is why it is named 'its stream fish run stream' [1-9, Kuslina Ck, locally 'Horse Ck'].

Figure 1-3. Ahtna houses at Eskilida Creek in about 1903.

This village site and the Fox Creek site [1-5] are two of the earliest recorded Ahtna place names, written by Dmitri Tarkhanov in 1797 as "Cheltekekat" and "Takekat" (Black 2008:71).

Photo from Geoff Bleakely, Wrangell-St. Elias National Park.

Yet kanii yełdu' Tats'esghi'aaden dae' hwdi'aan.
/There the next place upstream is called 'where a rock is in the water' [1-10, rock on E bank N of Saghani Na'].
3:46
Tats'esghi'aa yii ts'es tuu yii taghi'aayi gha su.
/'Where a rock is in the water', this is due to the fact that a rock is extending into the water there.

Tuu yii taghi'aa xunt'ae. Yii gha' su Tats'esghi'aaden dae' hwdi'aande yet.
/It seems to stick into the water there. That is why it is called 'where a rock is in the water'.

Big, bigdze' kayax kughile' yet ts'utsae du'.
/There was a big village there long ago.

Yet kanaa 'unii yełdu' Nic'anilen den dae' hwdi'aan.
/Then the next place across the way and upstream is called 'where current flows out from shore' [1-11, on W bank opposite Liberty Ck]
4:00
Nic'anilen den dae' tuu 'use su 'utggu nekedełeni, dansedze' nic'ana'idlen dze'.
/At 'where current flows out from shore', thus the water flows back out above there, then from out in midriver there it flows on back.

Yii gha' su Nic'anilen de dae' hwdi'aan de. You see, Nic'anilen de.
/This is why the place is called 'where current flows out from shore'. You see, 'where current flows out from shore'.

Yet kanii yak'a yet kanii xu yak'a Tselt'ogh Na'.
/There across, right across there is 'fractured rock creek' [1-12, Liberty Ck].

Now today means k'adii dzaene whiteman Liberty Falls dae' konii.
/   today the white man calls it Liberty Falls.

Liberty Falls dae' hwdi'aan k'adii du'.
/Now it is called Liberty Falls.

Yet ts'utsae yełdu' Tselt'ogh Na', yidi gha hwdi'aan, yic'a tsel, it's kind a funny name, you know Tselt'ogh Na'.[3]
/There long ago it was called 'fractured rock creek', for whatever reason it is so called, it is kind of a funny name, you know.

Dghilaay yits'itsiidgho'ł, dghilaay yits'itsiidgho'ł yii gha' su, you know ts'itsiidghotl'i naghitaeł xunt'ae łu c'a dighit'e' tuu, you know.
/The mountain has rock broken out from within, the mountain has rock broken out from within, that is why, rock is broken out from within and water rushes down.

Yii gha' su Tselt'ogh Na' dae' hwdi'aan, that' why, that's only that.
/That is why it is called thus, 'fractured rock creek', that is why.

It's kind a funny name all right. Tuu naghitaeł, see just right, Tselt'ogh Na'.
/It is kind of a funny name. The water rushes down, see just right, 'fractured rock creek'.

Duugh yet duugh yet kanii yełdu' yii gha' hwdi'aani.
/There next upstream a place is called for a reason.

Duu yae' kanaat Sdates, Sdates, dae' konii.
/Around there across the way is called 'peninsula portage' [1-13, site on point below Lower Tonsina].

Yełdu' Sdates tighita'i dae' su kanii. Sdates tighita' you know.
/There at 'peninsula portage' there was a trail upstream. 'Peninsula portage' was a trail, you know.

Kayax kughile'. Sdates tighita', yii gha' su Sdates dae' hwdi'aan.
/There used to be a village. 'Peninisula portage' had a trail, that is why it is called 'peninisula portage'.
5:18

---

[3] Jim may be thinking that **tsel** is a pun on the word 'rectum.'

Cu yet kanii yełdu' Ts'es K'et, Ts'es K'et cu yet cu dae' hwdi'aan, Ts'es K'et.
/Also the next place upstream is 'on the rock', another place called 'on the rock', 'on the rock' [1-14, bluff on W bank above Tonsina R].

Cu yełdu' ts'utsae nic'a'iltsiini dae' hdghiniixi, nic'a'iltsiini yii kiik'e ciisi t'ghił'aen'.
/Also they say that long ago there was a dipnetting dock, that dipnet dock upon it they used the dipnet.

You know, kiik'e ciisi tah uk'e t'ghal'aen'.
/You know, upon it a dipnet was used.

Łuk'ae uts'en 'ootnes.
/Salmon is taken from there.

Yii gha' su tats'es. . . . Ts'es K'et dae' hwniide. Ts'es K'et. Ts'es K'et they stand up, you know.
/That is why it is called (f.s.) 'on the rock'. 'On the rock, they stand on the rock, you know.

Ts'es K'et they stand up, ts'es k'et nkadzen xu ciisi ghił'iixi.
/At 'on the rock', they would stand at 'on the rock' and they would use a dipnet.
5:53
Yii su 'adetniiyi, that's why Ts'es K'et dae' hwdi'aande.
/That is how it is named, that is why it is called 'on the rock'.

Duu yet kanii yełdu' Nak'ay'taande dae' hwdetnii. Nak'ay'taande.
/There the next place upstream is called 'where willow extends', 'where willow extends' [1-15, on W bank below Bes Cene].
6:04
Yełdu' that bes 'unggu neketnghezdlaa xunt'ae you know bes neketnghezdlaa xunt'ae.
/Then there a bank curves to the upland, the bank curves around.

Yii yii t'aa nakustaan xu tkut'ae you know.
/Within this is a flat area, you know.

Nakustaan xu tkut'ae de yihwts'e k'ay' nastaan xunt'ae.
/Where there is the flat area there is a grove of willows.

Yii t'aa kayax kughile'i gha su, Nak'ay'taande dae' hwdi'aan Nak'ay'taande.
/Within there there was a village, named 'where willow extends', 'where willow extends'.

That's many people, many people raised there, you know.

Koht'aene nezyaande su.
/Many people were raised there.

Yet su 'adii kenii Nak'ay'taande.
/There now they call the place 'where willow extends'.

Deya'a kanii du' Bes Cene, Bes Cene yet yak'a k'adii bes, you know, 'uyggu hwt'aa yae' tighita' xunt'ae [1-16, Liebistag Village, Riverstag].[4]
/Right there next upstream is 'base of the riverbank, 'base of the riverbank' right now there is a bank and down below beneath there was a trail.

Hwt'aa yae' tighita', yet 'udaa'a ts'en yełdu' kayax kughile', you know.
/Beneath there there was a trail and on the downstream side there was a village, you know.
6:48
Yii gha' su Bes Cene hwdi'aan, Bes Cene. That's only that see.
/That is why it is 'base of the riverbank', 'base of the riverbank'.

Ut'aa yae' tighita'.
/Beneath there there was a trail that way.

Kadaa' yet kayax kughile' yet ye yii yet ya'a yii gha' su Bes Cene.
/Downstream there was the village, right there, that is why it is 'base of the riverbank'.
7:01
Should be little more something there, but that can't do it.

So Bes Cene that's far as he hwdi'aan.
/'Base of the riverbank' is just what it is called.

Yet dae' ye hwdi'aan kaniit cu ya kanii yełdu' Cetl'e's Cae'e, Cetl'e's Cae'e dae' hwdi'aan.
/From the place that is so called, next upstream then is 'sweet gale (plant) mouth', it is named 'sweet gale mouth' [1-17, site at creek N of Bese Cene].

Yełdu' 'adii naxu willow, different kind a willow, you know.
/There now are some types of willow plant, you know.

Yii dae' tsiis kedetsesi k'ent'ae, you know. Yii su 'adetnii.
/This plant looks like it is decorated with ochre. Thus it is said.

Tsiis kedetses xunt'ae you know. Different kind a willow.
/Like it is decorated with ochre, you know.

_____

[4] The early historic name Leibistag was recorded by Allen in 1885 here (see Figure 0-4). Sometimes it is pronounced 'Riverstag.' As I recall, Andy Brown did not know the origin of this word, **Bes Cene** being his home village.

Yii yet i'dilaeni yii gha' su, yii gha' su Cetl'e's Cae'e dae' hwdi'aande. Cetl'e's Cae'e.
/This is there and due to that, due to that it is named 'sweet gale (plant) mouth', 'sweet gale (plant) mouth'.
7:40
De yet kadaa'a kanaat yełdu' Sdaghaay dae' kenii.
/Then downstream and across the way they calls 'along the point'.

Sdaghaay yet c'a 'adii sdaa 'unset nic'akuni'aa xunt'ae you know.
/At 'along the point' there now a point goes far out in midriver from shore [1-18, point on E bank, N of Chestaslina].

Sdaa uk'e kayax kughile' see.
/On the point there was a village.

'Utggu kayax kughile' xu' ghile'. Yii su Sdaghaay dae' hwdi'aan.
/In the area above there was a village and so it was named 'along the point'.

Sdaghaay 'aden yen du' Sdaghaay Denen yen su yet ghida'en gha yet Sdaghaay Denen.
/At 'along the point' by there 'the person of along the point' [third chief title mentioned] stayed, so he was called 'the person of along the point'.

Yet ghat sdaa yet cu sden, that's name, yet su c'a xona u'uze dilaen, Sdaghaay, Sdaghaay dae' hwdi'aan.
/There is a point there, then there is the separate name, 'along the point', it's called 'along the point'.
8:15
Deyet kanii łdu' Ts'iyit'aen' Tayene' dae' konii, Ts'iyit'aen' Tayene'.
/There the next place upstream is called 'directly visible straight stretch', 'directly visible straight stretch' [1-19, stretch of river above C'anuu Cii].

Sii yii łdu' k'adii 'udaa' 'unaat ts'abael ghenaats'e' ts'eyaas, you know.
/As for me nowadays we go downstream, across there, we come out across from some spruce.

Now anyplace you know, highway cu, you know, some place straight, some place crooked, you know.

Yet yihwts'en yi ghen du' sii ghii ghenaats'en ts'its'eyaasen.
/There from there we come out across from the bend.

(We) went a long ways straight, you know.

Xu' su yii gha' hwdi'aani, yet Ts'iyit'aen' Tayene' you see a long ways.
/That is why it is named, there at 'directly visible straight stretch' you can see a long ways.

Ts'iyit'aen' Tayene' yii gha' hwdi'aan.
/'Directly visible straight stretch' is named due to that.

Yet kanii łdu' yet kanii yełdu', Gguux Hwts'iniyaa den, Gguux Hwts'iniyaa den dae' hwdi'aan see.
/From there the next place upstream is 'where a monster emerged, it is named 'where a monster emerged' [1-20, flat on W bank, Lincoln's fish camp].
9:00
Yełdu 'adii, I don't know, sii cu su datnii de su.
/Well then, I don't know, but this is what is said of the place.

You know, gguux, nen' t'aa gguux hwts'iniyaa den, dae' konii.
/You know, a creature, a creature came out from under the ground, it is said.

Kiinghił'aen' dahwniide, gguux nen' t'aax ts'iniyaa, cu łu, maybe I don't know, maybe there was something, I guess.
/It is said that they saw it there, a creature came out from under the ground, so maybe that is so, I don't know.

Really name is Gguux Hwts'iniyaa den, yii gha' su Gguux Hwts'iniyaa den hwdi'aan.
/The real name is 'where a monster emerged', and that is why it is named 'where a monster emerged'.

You can see that today you know, maybe you think that way too, I don't know.

Xu' c'a su Gguux Hwts'iniyaa den dae' hwdi'aani.
/So that is named 'where a monster emerged'.

Yet kanaa ye kanaa 'unii yak'a xona Tsedi Kulaen den dae' hwdi'aan see. Tsedi Kulaen den.
/There on the other side and upstream is called 'where copper exists', 'where copper exists' [1-21, Copper Village on E bank].

Cu yełdu' 'adii book gha c'a k'adan'a n'eł nahwghelnic, Tsedi Kulaen, tsedi udghats'ini'aa xunt'aeyi.[5]
/I just told you about that place for a book project, 'where copper exist', some copper had been protruding outward.
9:49
Kanay'nulaełi gha dyilaak, tadalnen.
/As he [an Ahtna man] tried to tie a rope around it, it fell into the water.

Duu yii gha' su Tsedi Kulaende dae' hwdi'aan, Tsedi Kulaen, yii tadalnen.
/So due to this it is named 'where copper exists', it fell in the water there.

Yii gha' Tsedi Kulaende dae', tsedi kulaen dae' hwdi'aan. Tsedi Kulaen Denen.
/That is why it is named 'where copper exists'. [Here was] the 'person of copper exists' [fourth chief title mentioned].
10:02
Cu kanii yak'a T'aghes Ciit. T'aghes Ciit dae' hwdi'aan see.
/The very next place upstream there is named 'cottonwood point' [1-22, point on W bank below Dadina R].

---

[5] Jim McKinley has told the story of the discovery of copper by the Ahtna at this location (Tuttle and Kari, to appear).

Figure 1-4. Caches on the west bank of the Copper River opposite the mouth of Chestaslina River (Ts'itazdlen Na'), near the village of Sdaghaay [1-18] on the east bank.

Photo by A. C. Spencer, July 19, 1900. Courtesy of U.S. Geological Survey. Courtesy of Bill Simeone, Alaska Department of Fish and Game.

T'aghes Ciit yełdu' 'adii łu t'aghes i'dilaen, you know, 'unset t'aghes nic'ani'aa xu tkut'ae, you know.
/At 'cottonwood point' now there is cottonwood, and cottonwood extends out in the open from shore there.

Yii t'aghes 'unse uyii ucii yet kayax kughile'i.
/There out in the open on the end of the cottonwood grove was a village.

Yii gha' T'aghes Ciit dae' hwdi'aan. T'aghes Ciit dae' hwdi'aan.
/That is why it is named 'cottonwood point'. It is named 'cottonwood point'.

Xu' łdu' 'unae' xu' łdu' let's see what's next, 'unae' yet kanaa kanii yak'a xona Hwt'aa Kughi'aaden dae' konii.
/Next on upstream, let's see what is next, upstream and the next place across and upstream then is called 'where an area extends below a place'.

Yii łdu' Hwt'aa Kughi'aaden means 'utggu bes dae' neketngezdlaa xu tkut'ae, you know.
/That 'where an area extends below a place' means that in the area up above a terrace curves around in that area, you know [1-23, bar on W bank below Dadina R].

10:48
Yii t'aa ts'abael ut'aa nastaan xunt'ae ut'aa yighi'aayi k'ent'ae.
/Beneath here, a grove of spruce extends beneath here, and it is as if it [the spruce] is something is extending below it.

Just like ut'aa yighi'aayi k'ent'ae.
/It is as if (the spruce) it is something that is extending below a place.

Yii gha' su Hwt'aa Kughi'aaden dae' hwdi'aan.
/That is why it is named 'where an area extends below a place'.

Hwt'aa Kughi'aaden that's kayax kughile', yet kayax kughile'.
/At 'where an area extends below a place' there was a village, a village was there.

Duu yet kanaa 'uniit yak'a xona Hwdaadi Na' dae' nii, yełdu'.
/Across from there and upstream next is 'downriver river', thus is said [1-24, Dadina R].
11:08
'Unggu, 'utggu K'ełt'aeni hwts'idiniłeni.
/In the area upland and above this flows out and down from K'eł t'aeni [1-25, Wrangell Mountains].

Yii dadaadze' kedełeni Yii Ts'itu' yii Ts'itu' yitaghił'aa xunt'ae you know.[6]
/It flows from the downstream to 'major river', the water flows on into 'major river' [1-6, Copper R (alternate name)].

Yii gha' su Hwdaadi Na' dae' hwdi'aan.
/So this is why it is named 'downriver river'.
11:20
Dii yet kanaa 'unii ya'a łdu' Utl'aa Ts'esz'aani, dae' konii,
/There then across the way and upstream is called 'its headwaters has a rock' [1-26, stream on W bank above Dadina R].

Utl'aa Ts'esz'aani, Utl'aa Ts'esz'aani yii łdu' you utl'aat little na' yii cu.
/'Its headwaters has a rock', 'its headwaters has a rock', you know, at the head of this little stream too.

Just little thing, you know, not too much big.

Nggu utl'aa łdu' cu ts'es dini'aa, you know.
/Upland at its headwaters a boulder extends.

Utl'aa ts'es dini'aa xunt'ae, yii gha' su cu yełdu' Utl'aa Ts'esz'aani hdae' udetnii.
/At its headwaters a boulder is extending and this is why then it is called 'its headwaters has a rock'.

---

[6] **Ts'itu'**, literally 'straight water,' is an alternate name for the Copper River, used by Jim McKinley here instead of **'Atna'**. The same name can be used for the Susitna and Tanana rivers.

Yii kayax kughile', yii na' k'a, kayax, Utl'aa Ts'esz'aani Cae'e.
/This used to be a village, at that stream the village is 'mouth of its headwaters has a rock' [1-27, mouth of stream on W bank above Dadina R].

Yet yii kanaat yak'a Naak'e, Naak'e dae' hwdi'aan.
/Across from there is 'on the bar', it is named 'on the bar'.

Naak'e yełdu' k'adii k'adii Naak'e Naak'e kughile', you know.
/'On the bar', there now is 'on the bar', 'on the bar' used to be there [1-28, bar 2 mi below Nadina R].
12:00
'Unsu Naak'e k'ay' ggaay nastaan xu, t'aghes little nastaani yii tah kayax kughile', you know.
/Out in the open at 'on the bar' there are stands of small willows and stands of little cottonwood and there was a village there too, you know.

Yii gha' su Naak'e dae' hwdi'aan, Naak'e.
/That is why it is named 'on the bar', 'on the bar'.

Basiili k'adii 'asnii yen ak'ae kughile' yet, Basiili ghida'.
/Chief Bacili, who just I spoke of, that was his location, that is where he stayed.

De yet kanaat kanii k'a xona Nic'akuni'aaden dae' hwdi'aan.
/Across the way and upstream of there is named 'where area extends out from shore' [1-29, site on E bank 1 mi above Nadina R].

Nic'akuni'aaden. Yełdu' 'unse sdaa nic'akuni'aa.
/At 'where area extends out from shore' there a point extends in the open way off from shore.

'Unaa 'unaa bes 'unaaxe hwnez'aani yii c'axuni'aa xu xunt'ae, 'unaaxe nekedełen, 'Atna', you know.
/Across there, across there the bank goes up and it extends on away (from shore), the Copper River flows around and across there, you know.

Yii gha' su Nic'akuni'aaden dae' hwdi'aan, see.
/This is why it is named 'where area extends out from shore'.

You see pretty near he touch that bank.

Yii gha' su Nic'akuni'aaden dae' hwdi'aan.
/This is why it is named 'where area extends out from shore'.
12:43
Yet kanii yak'a xona Nige' Kulaen Nuu T'ax, Nige' Kulaen Nuu T'ax dae' hwdi'aan.
/The next place right upstream is 'behind silverberries exist island', it is named 'behind silverberries exist island' [1-30, islands near E bank near site, 5 mi above Nadina R].

Duu yii łdu' k'adii nigige', den gige' dae' konii.
/Here now is this berry, a berry that we usually call 'land berry'.

Den gige' means I guess you can't I don't know xay tah nelggayi xunt'aeyi yi c'encaes hkedghiłt'e' ts'utsae, you know.
/The land berry (silverberry) is a berry that is white in the winter, and long ago they used it for a berry jam.
13:08
Dengige', you know.
/The silverberry.

Yii gha' su Nige' Kulaen Nuu T'ax dae' hwdi'aan.
/That is why it is named 'behind silverberries exist island'.

Yi yii i'nilaeni, yii gha' Nige' Kulaen Nuu T'ax dae' hwdi'aan see.
/This (berry) is there, and that is why it is named 'behind silverberries exist island'.

Duu yet kanii k'a xona, yet kanii xona bes . . . , yet kanii xona T'aghes Tah, T'aghes Tah dae' hwdi'aan see.
/Upstream from there, the next place upstream, the next place upstream there, is 'among the cottonwoods', it is named 'among the cottonwoods'.

T'aghes Tah yet stsiye ghida'.
/My grandfather stayed there at 'among the cottonwoods' [1-31, Wood Camp].
13:44
Yet cu t'aghes dilaen dze' ye tah kayax kughile'i gha su yet T'aghes Tah dae' hwdi'aan, T'aghes Tah.
/There is also cottonwood there, and used to be a village there, and due to that it was named 'among the cottonwoods', 'among the cottonwoods'.

Yet kanaaxe kanaaxe cu bes nez'aan, c'a yet'aa cu danaa'a ts'en yełdu' Naniłts'elyaak Bese' dae' hwdi'aan.
/Right across, in the area across is another bank on the other side, and that is named 'we did something to each other riverbank'.

Naniłts'elyaak Bese' duu yiłdu' koht'aene yenidan'a tah koht'aene inaniłzelghaeni.
/At 'we did something to each other riverbank', then some people, in ancient times some people killed one another there.

Yii gha' su Naniłts'elyaak Bese' dae' dae' hwdi'aan see.
/That is why is it named thus, 'we did something to each other riverbank' [1-32, bank opposite Wood Camp].

Yet c'a kayax kughile'i gha su Naniłts'elyaax Bese' dae' hwdi'aan see.
/There too there was a village at the place named 'we did something to each other riverbank'.

[Where] people [were] living that's all I talk about now see.[7]

---

[7] Jim emphasizes that he is not calling all of the named Ahtna places, but that he is highlighting the locations of Ahtna settlements.

Yet kanaat yak'a Tl'atibese', Tl'atibese', Tl'atina' Bese' ghenaat.
/There on the other side is 'headwaters bank', 'headwaters bank', 'headwaters river bank' (Klutina R) is on the other side [1-33, bluff on W bank, S of Klutina mouth].
14:30
Yii cu 'unaat ndahwdi'aan kangge' cu little ways two place hwdi'aan. Tl'atibese'.
/There is another place across from there named this way, upland a little way, two places have the name 'rear bank' [the upstream Tl'atibese', 1-34, above the mouth of Klutina R].

Tl'atibese' yii łdu' k'adii koht'aene hdaghalts'e' sii snaghał koht'aene hdaghalts'e', you know.
/Now people used to live at 'rear bank', in my time people lived there.

Tl'atibese' yet kanaat ya'a Utay'laxi Na', Tay'laxi Na', yet łdu' łuk'ae una' talaxi yii gha' su xu' hwdi'aani, łuk'ae una' talax.
/From 'headwaters bank' right there across the way is 'fish-run stream', 'fish-run stream', there the salmon run in the creek, and that is why it is named, salmon run in the creek [1-35, creek from E half mile below Klutina R mouth].

Yet Una' Tay'laxi Na'.
/That is 'its stream fish run-stream'.
14:58
Yet kanii c'a xona Tl'aticae'e
/The next place upstream is 'rear mouth' [1-36, mouth of Klutina R]

Tl'atina'.
/and 'rear river' [1-37, Klutina R].

K'etl'aa tina' 'unggase' gheniit ts'ediniłen, you know.
/From the headwaters of the river, it flows out from upland at the upstream end, you know.

Yii gha' su Tl'atina' dae' hwdi'aan.
/That is why it is named 'rear river'.

Ye ye kanaat ts'e xona Kolghosi Na' kolghosi that means lots of little things.
/Across from there is 'foamy stream'. Kolghosi means lots of little things.

Niłc'aaydze' brush dilaeni su Kolghosi Na' dae' hwdi'aan.[8]
/Various kinds of brush are there and it is called 'foamy stream' [1-38, Klawasi Ck].

Niłc'aaydze' cu lots of things, lots of different kind a brush, yii gha' su Kolghosi Na' dae' hwdi'aan.
/Various things, different kinds of brush are there and that is why it is called 'foamy stream'.

---

[8] Sometimes translated as 'boiling or foamy water stream'. The names Klawasi and Klutina were first mapped in 1885 by Allen (see Figures 0-4 and 1-6).

Figure 1-5. The Lower Klawasi mud volcano.

The sulfurous pond by the "Mud Volcano" is called **Natu' Bene'** 'salt water lake' and the a couple of streams draining to the Copper River are called **Natu' Kaghił'aa** 'where salt water flows up and out' [1-41].

Photo by geologist Suzanne McCarthy of Prince William Sound Community College, who is doing a long-term study of this area.

15:39
Yak'a k'adii yet kanaat ts'e xona Tes K'et, Tes K'et dae' hwdi'aan.
/Right there across the way then is 'on the hill', it is called 'on the hill'.

Yii cu kayax kughile'.
/That was another village [1-39, bluff at Copper Center airport].

Right across there this side of mission where we live now.

Tes K'et dae' hwdi'aan.
/It is called 'on the hill'.

Yet kanii xona 'adii naene tnaey hdelts'iidi yełdu' k'adii cu 'adii xu' hwdi'aan Tatsen Na' dae' hwdi'aan.
/Across from there where we people now live, there now is named 'smelly water creek' [1-40, Yetna Ck].

'Adii 'sdelts'iide Tatsen Na'.
/Now where the people live is 'smelly water creek'.
16:10
Yet koht'aene yełdu' tatsen yi du' water, you now, tuu tuu cuts'endze' neltsiini xu' tkut'ae something,
/There in the language is that smelly water, it is a water that looks different.

You can't drink that water. 'Ele' c'a see.
/   No.

Salt tuu uhwnelnes the way it is, natu' koniiyii hwnelnes, yii gha' su tatsen dae' hwdi'aan.
/It tastes of salt, the way it is it tastes of salt, that is why it is called 'smelly water'.

Tatsen Na' dae' hwdi'aan.
/It is called 'smelly water creek'.
16:31
Yet kanaa 'unii xona tu' tuu yii Natu' Kaghił'aa, you know, kat Natu' Kaghił'aa tkut'ae.
/Across from there and upstream then is, in the water is 'salt water flows up', as salt water is flowing up from below [1-41, creek from E below Nay'dliisdini'aaden; see Figures 1-5 and 1-6].

Duu yet kanaat ya'a xona Nay'dliisdini'aaden, Nay'dliisdini'aaden dae' hwdi'aan.
/Right across from there then is 'where songs extend across', the place is called 'where songs extend across' [1-42, sites on both sides of Copper R near Silver Spring].

You see Nay'dliisdini'aaden yet maybe two thousand years ago cu su kughile' ts'inaandze' kentighita' dae' konii.
/You see at 'where songs extend across', maybe two thousand years ago there were trails coming straight across (the river) here.

Kon'tah teghiyaas ne yene yene 'ele' c'a na'ittl'uul dze' 'unaaxe kon'tah ghidaax.
/The ones who were going visiting, they would not even be dressed (for the cold), as they just would visit across the way.

He come back, dae' kenii see, yii gha' hwdi'aani su Nay'dliisdini'aaden.
/He would come back, they say, and due to that it is named 'where songs extend across'.

You can hear hwt'ae' 'unaa 'unaaxe c'ehdelyaesi yii c'a hdighits'ak.
/They can hear them singing from across the way.
17:12
Dae' hwts'en ye niłdiłna you know, yii gha' hwdi'aani su Nay'dliisdini'aaden.
/Thus due to that (sound of songs) going back and forth, is why it is called 'where songs extend across'.

'Adii 'adii whiteman nec'a' kuzniic ghayet c'a, you know.
/Now the white man has taken this place from us, you know.

Yet kanii ya'a xona Natu' Kaghił'aaden dae' hwdi'aan. Yii c'a Natu' Kaghił'aa.
/There the next place upstream is called 'saltwater flows up', there is [where] the saltwater flows up [1-43, creek with Mineral Spring].

Yet kanaat ts'e xona Tezdlen Cae'e, Tezdlen Na' Bese'.
/The next place on the other side then is 'mouth of swift current', and 'swift current stream bank' [1-44, mouth of Tazlina R].

Gha yuu Dadaa'ats'en, Tezdlen Na' Bese', yet da 'Udaa'ats'en Tezdlen Na' Bese'.
/By there is 'downstream side swift current stream bank', and there is 'downstream side swift current stream bank' [1-45, Tazlina Terrace, bank on W below mouth of Tazlina R].

Yet kanii yak'a xona Tezdlen Cae'e, ucae'e Tezdlen ucae'e dae'.
/There the next place upstream is 'mouth of swift current', the mouth of 'swift current' [river].

C'a kanii ya'a xona yet Tezdlen Bese', Tezdlen Bese' dae' hwdi'aan see.
/So the next place upstream is called 'swift current bank', 'swift current bank' [1-46, bank on W above mouth of Tazlina R].
18:05
Kanaa yii su long time ago nenatseh tah c'akunizet yet, man drown dyaak dze'
/Across there before us long ago a person drowned there, he drowned and

yii gha' hwdi'aani su, Tezdlen Bese' see, Tezdlen Bese'.
/due to that it is named 'swift current bank', 'swift current bank'.

Now this k'adii 'Atna' ya'a k'adii Frank Stickwan yet hdelts'ii de xu su cu 'ele' kayax hwghilaelde ts'utsae. you know.
/Now here where Frank Stickwan [Stickwan fish camp] stays, there was no village long ago.

That's why no name there nothing.
18:30
See that flat, that big k'adii su xu' kayax kughile', whiteman down there, I guess.
/You see, that flat there is a village now, white man down there now, I guess.

Kayax kuzdlaen, you know, no name there.
/Where the (Tazlina) village is there is no name.

Ya xona 'unii yak'a xona Sday'dinaesi gha, yet c'a tnaey hdaghalts'e' you know Sday'dinaesi.
/There then upstream is is 'long point', here too people had lived, you know, 'long point' [1-47, point on W bank at Glennallen].

That means sdaa, long sdaa you know, 'unset sdaa dighiłnaesi yii gha' su Sday'dinaesi gha dae' hwdi'aande, Sday'dinaesi you see.

/That is a peninsula, a peninsula goes way out [in the Copper R], and that is why the place is called 'long point', 'long point'.
18:57
Yet kanii yak'a Latsibese' Caek'e, that's little ways from there, Latsibese' Caek'e.[9]
/There the next place upstream is 'Latsi bank mouth', that is a little way from there, 'hand head bank mouth' [1-48, mouth of Dry Ck, Dry Creek Village].

Martha Jackson's daddy yet ghida' you know, Latsibese' Cae'e.
/Martha Jackson's father lived there you know, 'hand head bank mouth'.

Dii yet kanaa ya'a kaniit yak'a xona Tes K'et.
/There next across and right upstream is 'on the hill'.

Tes K'et tnaey hdaghalts'e' fish are there łdu'.
/At 'on the hill' people lived on the hill, fish are there.

Tnaey hdaghalts'e' you know, dae' yii gha' su Tes K'et dae' hwdi'aan, nice looking place there.
/People lived there and that is how it is called 'on the hill', it is a nice looking place there [1-49 bank on Copper R above Dry Ck].
19:32
De yet kaniit yak'a, yet kanii yak'a xona Tatsengha, Tatsengha yet kayax kughile' yet, Tatsengha.
/The next place upstream, the next place upstream then is 'by smelly-water', 'by smelly-water' was a village, 'by smelly-water' [1-50, site 1½ mi above Dry Ck mouth].

I remember that, yet cu snaghał 'adii tnaey hdelts'iix xu ko'ii kusya' Tatsengha.
/I remember that, in my time people stayed there by the time I was conscious (in early childhood) at 'by smelly-water'.

Dae' yak'a yii gha hwdi'aan su Tatsengha.
/This is how it is called, 'by smelly-water'.

Tuu kaghił'aa you know. Natu' kaghił'aa yak'a yii gha' hwdi'aani su Tatsengha.
/Water flows up, a salty water flows up and that is how it is named 'by smelly-water'.
20:00
De yet kanii xona C'ulc'e Cae'e, C'ulc'ena' dae' hwniigha,
/There the next place upstream is 'mouth of C'ulc'e', and what is called 'C'ulc'e river',

ucae'e C'ulc'e Cae'e dae' hwdi'aan. [1-51, mouth of Gulkana R; 1-52, Gulkana R]
/its mouth is called 'mouth of C'ulc'e'.

Xu baygha little ways from there C'ulc'e Cae'e dae' hwdi'aan.

---

[9] Here Jim uses the Central Ahtna dialect form **caek'e**, whereas he had been using his Lower Ahtna dialect **cae'e** for 'mouth'.

/A little ways from it, it is called 'mouth of C'ulc'e.

Yiłdu' C'ulc'ena', I don't know what really means.
/Then 'C'ulc'e river', I don't know what that really means.

'Ele' I never mention that thing.
/Not   .

C'ulc'ena' C'ulc'ena'.[10] I don't know something. Just xu'a dghut'e'de, you know.
/   That is just the way it (the name) is.
20:25
Yet kanaa 'uniit xona łuk'ae gha hdelts'iix den, Łuk'ae Gha Hdelts'iixden.
/Across from there them is where they stay for fish, 'where they stay for salmon' [1-53, fish camp above mouth of Gulkana R].

That means fish or salmon pick 'em up there. Łuk'ae gha hdelts'ii, yet tnaey hdaghalts'e' you know yet łuk'ae hdelts'ii.
/   They stay for fish, and there people used to stay.
22:05
Dii yehwts'en xona Ggax Kucae'e, Ggax Kucae'e, yet Ggax Kucae'e yak'a.
/Then from there is 'rabbit area mouth', right there is 'rabbit area mouth'.

Yełdu' k'adii ggax ts'utsae una' ghile'i yii gha' I think, yii gha' udi'aan, Ggax Kucae'e dae' hwdi'aani.
/Long ago at this stream there were rabbits, that is why it is so named, it is called 'rabbit area mouth' [1-54, mouth of Gakona R].

Yet kanii Ggax Kucae'e, Ggax Kuna'.
/The next place upstream of 'rabbit area mouth' is 'rabbit area river' [1-55, Gakona R].
21:05
Ucaek'e tnaey hdaghalts'e' you know; ucaek'e tnaey hdaghalts'e' yet.
/At its mouth people lived, at its mouth there people lived.

Yet kanii k'a Tazaan Nuu' Tah, Tazaan Nuu' Tah dae' hwdi'aan.
/Upstream of there is 'island in clear area', it is named 'island in clear area' [1-56, Old Gakona, Gene homesite, below 5 mi. hill].

Yełdu' that's other language you know that tazaan I think that's up the line language, tazaan means 'clear', I think, clear country long, that time no tree much, I guess, not much, I guess.
21:35
Yii gha' c'a su, today you can't see, lotsa tree there.
/However today you can't see, there are lots of trees there.

---

[10] The name for Gulkana River is associated with the verb **c'el** 'to tear', but the name is irregular in structure. Note that 'Tonkina', one of the few names via Allen (1887:63) that is not recognized by Ahtna speakers, appears to be for the Gulkana River. See Figure 0-4.

Figure 1-6. Sketch maps are a useful technique for eliciting and mapping sequences of names. On this sketch map of the central Copper River and side streams drawn by Tenas Jack in Copper Center on July 12, 1982, Tenas mentioned 20 names, 13 of which are also in Jim McKinley's narrative, as indicated by the numbers. In discussions with well-traveled Ahtna speakers, reiteration and corroboration of Ahtna names are routine.

Tazaan Nuu' Tah that's what it mean, no nothing there, Tazaan Nuu' Tah, dae' hwdi'aan you know.
/'Island in clear area', that's what it means, it is clear, thus it is 'island in clear area'.

That far yet c'a kanii c'a xona Taltsogh Na'
/Right upstream of there is 'yellow-water creek' [1-57, Tulsona Ck]

Taltsogh Cae'e, Taltsogh Cae'e.
/and 'yellow-water mouth' [1-58, mouth of Tulsona Ck].

That means tuu neltsoghi yii gha',
/That means water is yellow that is why.

Tuu neltsoghi yits'inił'aa ts'e yii gha' su Taltsogh Cae'e dae' hwdi'aan.
/Yellow water flows out from there and that is why it is named 'yellow water mouth'.
22:08
Taltsogh Cae'e gha yet ka'aa c'a Taltsogh Bese' cu dae' hwdi'aan, Taltsogh Bese', that's bank there, Taltsogh Bese' dae' hwdi'aan.
/There over from 'yellow water mouth' is named 'yellow water bank', 'yellow water bank' has a bank there, and is so named [1-59, bluff on E side of Tulsona Ck].
22:18
A reordered and inserted segment follows. This is the only out-of-sequence place that Jim McKinley presented that day.
Begins 22:21
Yii si cu you know that Ts'uuni su ts'utsae little island you know, little place yii su yet ts'utsae koht'aene kaghaltsiin dae' koniiyi
/That place 'bump' long ago was a little island, long ago the people [a clan] were made at that place, So it is said [1-60, island with site on Copper River near 19–20 mi. of Tok Cutoff].

Uk'et koht'aene kaghaltsiin dae' hwnii.
/On it the people (a clan) were formed on it.

K'adii nanii Adam yene 'Ałts'e' tnaey ye kakaltsiin dae' konii, yii gha' su.
/Now upstream Adam [Sanford] has said that they Ałts'e' tnaey people were formed there, it is said.

That little things like that, I don't know, people just can't believe, you know. I don't know.

C'a k'ał'aa c'a su tkonii de, maybe k'ał'aa c'a cu du' konii someway.
/But what is said is be true, it may be true in some way.

Somebody gotta tell that one, you know.

Yehw koht'aene, koht'aene yet kaghaltsiin dae' konii.
/'There a tribe of people were formed there, it is said.

See yen yene k'adii naxu Adam [Sanford] Adam ne'eł k'adii, or 'adii sunghae dae' Old Charley kiidghine'en yen 'eł ghida'en 'eł
/Those, now Adam is with us or my older brother they called Old Charley [Sanford] said that and

yene yen 'adii u'eł snakaey ba'ane hdelts'ii, ucaek'e hdelts'iin.
/they are the children [of the clan] staying out beyond there, the ones who live at the river mouth (Chistochina people).

Yen itat'aex dghine'i, you know.
/That is according to what he had said, you know.

You see that's far as I know that thing, that Ts'uuni.

That's something that interest[ing], I don't know, maybe right c'a su kenii.

We don't know see. That's the story on that little island there.
23:30, end inserted segment at 23:32. Continue first segment.
Yet 'unae' c'a xona Sdacen Tayene', ucighi'aa, you know, Sdacen Tayene' means big long bar there.
/There upstream then is 'flat-point straight-stretch', a point extends, you know, 'flat-point straight-stretch' means a big long bar is there.

Yii gha' su Sdacen Tayene' dae' hwdi'aan.
/So that is why it is called 'flat-point straight-stretch' [1-61, straight stretch on Copper R near 20 mi.].

Then yet 'unae' yak'a xona Bany'didaasi Tayene'.
/Then past there and upstream then is 'something (caribou) crosses straight-stretch' [1-62, bar on Copper R near 29 mi.].

Banay'didaasi Tayene' that means caribou go across on there, caribou go across on there.
/'Something (caribou) crosses straight-stretch' means that caribou go across there.

Yii gha' su hwdi'aan, Bany'didaasi Tayene'.
/That is why it is called 'something (caribou) crosses straight-stretch'.

Bany'didaasi Tayene' dae' hwdi'aan.
/It is called 'something (caribou) crosses straight-stretch'.
24:06

Yii kadaat Bayn'didaasi Tayene' yihwts'en tah xona Tsiis Tl'edze' Caek'e.
/There next downstream at 'something (caribou) crosses straight-stretch' from there then is 'blue ochre mouth' [1-63, mouth of Chistochina R].

Tsiis Tl'edze' Caek'e, yii łdu' 'adii tsiis ts'utsae kiikae ninattsiisi.
/'Blue ochre mouth', there long ago they used that ochre to paint themselves.

Yii utl'aa i'dilaeni gha su Tsiis Tl'edze' Caek'e dae' hwdi'aan. Tsiis Tl'edze' Caek'e.
/Since it is available at its headwaters it is called 'blue ochre mouth'. 'Blue ochre mouth'.

Ts'utsae tsiis uke'sdghiłt'e', yii gha'.
/Long ago we used to use the ochre, due to that.

Kayax kughile' yet c'a, yet kaniit c'a xona Tes K'et, another one, another Tes K'et tnaey hdaghalts'e' yet.
/There was a village there, and upstream of there is 'on the hill', another place called 'on the hill' [1-64, hill on N bank of Copper R].

Tes K'et 'adii cu yet some hdelts'ii I think, Tes K'et.
/At 'on the hill' now some people still live there I think.
24:56
Kanii yet xona Tsedghaan Na', I don't really mean that one. Tsedghaan Na', I can't tell, up the line, I just can't don't know see, Tsedghaan Na'.
/Next upstream is 'moldy rock stream', I don't know what that means. Up the line (in the Upper Ahtna dialect) [1-65, creek into Copper R Indian River from S].

Adghaani ts'es, yii na' su dae'.
/A moldy rock, that stream is like that.

Kanaa'a xona Tay'slaex, Tay'slaexden, Tay'laxden, dae' another one again Tay'laxden.[11]
/Next across from there then is 'fish run', 'where fish run', another place by that name [1-66, possibly Łuk'ece'e Na'].
25:26
Yet kanaa yak'a xona Baa Łaedzi.
/Then right across from there then is a bank, 'grey soil'.

Baa Łaedzi is the last one from Banzanita, Baa Łaedzi.
/'Grey soil' is the last place before Banzanita.

That means bes too yii c'a, Baa Łaedzi, Baa Łaedzi.
/That too means 'grey soil' [1-67 bank on N side of Copper R 1 mi. above Slana].

Unen łaets łaets 'eł when wind wind time,
/On the hillside is soil,

when big wind you can see that, just like smoke.

---

[11] Jim may have substituted the name for Mabel Creek off the Slana River; the stream here is likely **Łuk'ece'e Na'.**

Baa Łaets, Baa Łaedzi yii gha' su Baa Łaedzi dae' hwdi'aan yet, Baa Łaets dae' hwdi'aan.
/'Grey soil', it is called 'grey soil' soil due to this, it is called 'grey soil'.

I think that as far as . . .
26:00
Kanii c'a xona, Stl'aa Caek'e dae' hwdi'aan, Stl'aa Caek'e,
/Next upstream then is called 'rear mouth', 'rear mouth' [1-68, mouth of Slana R].

Cu yii łdu' uyii takudesedi gha I think, deep water, uyiit, deep water, you know, Stl'aa Caek'e that's why hwdi'aan.
/That place is for its deep water, it is due to the deep water, you know, and that is why it is called 'rear mouth'.
26:22
Yihwts'en, Yet ts'en, yet kanii c'a xona Nataełde,
/From there then is Nataełde [1-69, 'roasted salmon place' Batzulnetas].

Nataełde that mean clear place there you know, nice looking place there, yii gha' su Nataeł de dae' hwdi'aan, that's far kayax kughile'.
/That means a clear place, nice looking place and that is why it is called Nataełde, there was a village there.

That's far as. . . . Yihwts'en xona, xungge' Sasluugge', yet katggat, Sasluugge' dae' konii.
/From there then upland is 'sas fish', above there the place is called 'sas fish' [1-70, Suslota village site].

Yiiłdu' Sasluugge' yii una' talaxi yii gha' su łuk'ae una' talaxi gha' Sasluugge'de dae' hwdi'aan see.
/At that 'sas fish' in that creek fish (a small sockeye salmon) run, and due to that type of salmon running it is named 'sas fish'.
27:00
Yet kangga k'a xona Tak'ez'aani, yiłdu', big tes ce'e yet nez'aani, yii su unekekusdaan xunt'ae you know.
/Then next upland is 'the one in the valley', there is a big hill there, and it has a ravine surrounding it, you know [1-71, two hills NE of Tuu Ts'eni, mountain on N side of Indian Pass].

Yii yi yi gha' su Tak'ez'aani dae' hwdi'aan see.
/That is why it is named 'the one in the valley'.

Yet kangga xona Mendaesde, Bendaesden dae' hwdi'aan Bendaes den.
/Next upland it 'shallows lake place', it is named 'shallows lake place', 'shallows lake place' [1-72, Mentasta].

See that's something that, yedi c'a gha' hwdi'aani yii c'a?
/Now why is that called this way?
Ends at 27:34

Part 2
Tl'atina' Ngge'
The Klutina River Drainage

Rec. Jan. 12, 1981 AT 23 or AT 5005; total: 17:47. Sound file chp1-2-jimmckinley.wav
Recorded in the same day as Part 1; the first 3:50 of this session has three place anecdotes that are not sequential travel narratives.

Starting point: Copper Center
Ending points: Valdez and the uppermost Tonsina River

Cu little bidze' nahwgholnic, you know.
/Let me tell a little bit more.

Tl'atina' Ngge' gha.
/About the 'rear water river uplands' [1-73, entire Klutina R drainage].

Tl'aticae'e dae' konii. yii ucae'e yegha ts'inił'aayi gha su.
/'Rear water mouth' [1-35, mouth of Klutina R] thus it is said. There at the mouth the current flows out by there.

Tl'aticae'e dae' konii de.
/Thus is said 'rear water mouth'.

Yet kanggat yełdu' Ts'ekul'uu'i Cae'e dae' konii.
/The next place upland of there then is 'one that washes out mouth' [1-75, stream from N, 1 mi. up Klutina R, on lower terrace], thus is said.

Yełdu' ts'ekul'uu' yeghi ts'es de, ts'es ighi tuu ts'e tuu tuu yii daagge' ts'ekul'uu'.
/So then they wash out, there rocks, rocks there due to the water, they wash out of there.

Tuu naghił'aayi daagge' ts'ekul'uu'. Yii gha' hwdi'aan, Ts'ekul'uu'i Cae'e.
/Due to the water flowing down, they wash out. Due to that it is named 'one that washes out mouth'.

:39
Ts'ekul'uu'i Cae'e su.
/That is 'one that washes out mouth'.

Ye kanggat łdu' Ba'ane Ts'ilaaggen Tak'adze' dae'.
/Then the next place upland is 'spring of someone killed him outside'[12] [1-75, site and spring on N side, 4 mi. up Klutina R].

---

[12] The location is not certain, but it must be on the lower terrace of the Klutina River.

Yen łdu' that's a person's name, you know.
/He then, that's a person's name, you know.

Ba'ane Ts'ilaaggen somebody kill him.
/'Spring of someone killed him outside'.

Way down to Ellamar some place you know. He's chief too that man. Yen su he make, he got camp there.
1:00
Yii gha' hwdi'aan su, Ba'ane Ts'ilaaggen Tak'adze'.
/That is what it is called 'spring of someone killed him outside'.

He got camp dae' kughisiin' yii gha' Ba'ane Ts'ilaaggen Tak'adze'.[13]
/He had a camp there, due to that thus is 'spring of someone killed him outside'.

Gha yet k'a ya'a łdu' cu Bes Ce'e dae', Bes Ce'e means ut'aa ut'aa ts'inił . . . ut'aats'e.
/There then right there also 'big bank' [1-76, bank on N side of Klutina R at 5 mi.], thus, 'big bank' beneath it there.[14]

Xu Tl'atina' 'udighił'aax xunt'ae, you know.
/The 'rear water river' flows past, it is, you know.

Xu' yae' just close way. Tl'atina' banił'aa, bank, you know.
/Thus it is close (the bank to the river), 'rear water river' flows up to it, you know.

Bank bank yii uze' su, bes, you know.
/It is named for a bank, a river bank, you know.
1:37

Yet kangga ya'a łdu' Tak'ats' Kaghił'aaden dae'.
/Next upland there is 'spring water flows up' [1-77, spring on Klutina R at 8 mi., near 'Kids' Farm'] thus.

Yet su cu miracle c'a tkut'aedze' yet.
/There a miracle [a remarkable event] happened there.

Tak'ats' Kaghił'aa yet. You know, bes niidze su Tak'ats' Kaghił'aa.
/'Spring water flows up' there. You know, from the upstream of the bank spring water flows up and out.

---

[13] **Ba'ane** 'outside' means that this person was killed when he was outside the Copper River Basin, in Ellamar on the coast. This is a rare Ahtna commemorative place name; the name is also of importance in that makes note of an incident that happened outside of Ahtna territory.

[14] In fact, there are three places on the Klutina River named 'big bank'; see 1-89 and 1-90.

Bes yits'iniłtaeł xunt'ae, yii gha' hwdi'aani su.
/It rushes out from within the bank, that is why it is so named.

Tak'ats' Kaghił'aade dae' hwdi'aan.
/'Spring water flows up' thus it is named.
2:01
Yet kangga'a łdu' Tl'atiditaande.
/There the next place upland then is 'where trail goes to head (of stream)' [1-78, N-S trail at 8.5 mi. going up esker toward Hudson L].

Tl'atiditaande,
/'where trail goes to head (of stream)',

that means old trail go down to the river.

Yii gha' hwdi'aani su.
/That is reason it is so named.

Deniigi Kaen'. Oh, I forgot the name. Little place there.
/'Moose lodge' [1-79, hill to N, up Tl'atiditaande, revised location from Kari 2008 #434]. I forgot to mention.

[I] don't mention that Deniigi Kaen'.
/I did not mention that 'moose lodge'.

Ye Katiztaan, Katiztaan.
/There is 'the trail ascends', 'the trail ascends' [1-80, incline/decline on trail at 13–13.5 mi., not in Kari 2008].

Hwnitinitaande dae' du' Hwnitinitaande dae' hwdi'aan dae'.
/'Where a trail terminates' [1-81, a N-S trail to Hudson L, at 8–9 mi.] thus then 'where a trail terminates' thus it is called.[15]
2:37

Yet hwts'en kungge' kanggat kudełdeyede ya'a.
/There from there to the uplands it is just a short distance to the uplands (to the area north toward Hudson L).

Sdaak'e Natinitaande dae', yet łdu' sdaa unse' ts'ini'aay yi k'e yi k'e natinitaan xu tkut'ae. Yii gha'.
/'Trail crosses the point' [1-82, trail crosses bend on Klutina R between 12 and 13 mi.] thus there then, a point extends on out, and on it, on it a trail goes across, is the way it is. That is why.

Yet yet kangga' ghanaay Ghanaay Zelghaende udetnii.
/There to the uplands caribou, 'where a caribou was killed' [1-83, hill and creek before Hudson L trail near 12 mi.], it is said.

---

[15] The old trail likely paralleled the current Hudson Lake all-terrain-vehicle trail several hundred yards to the north.

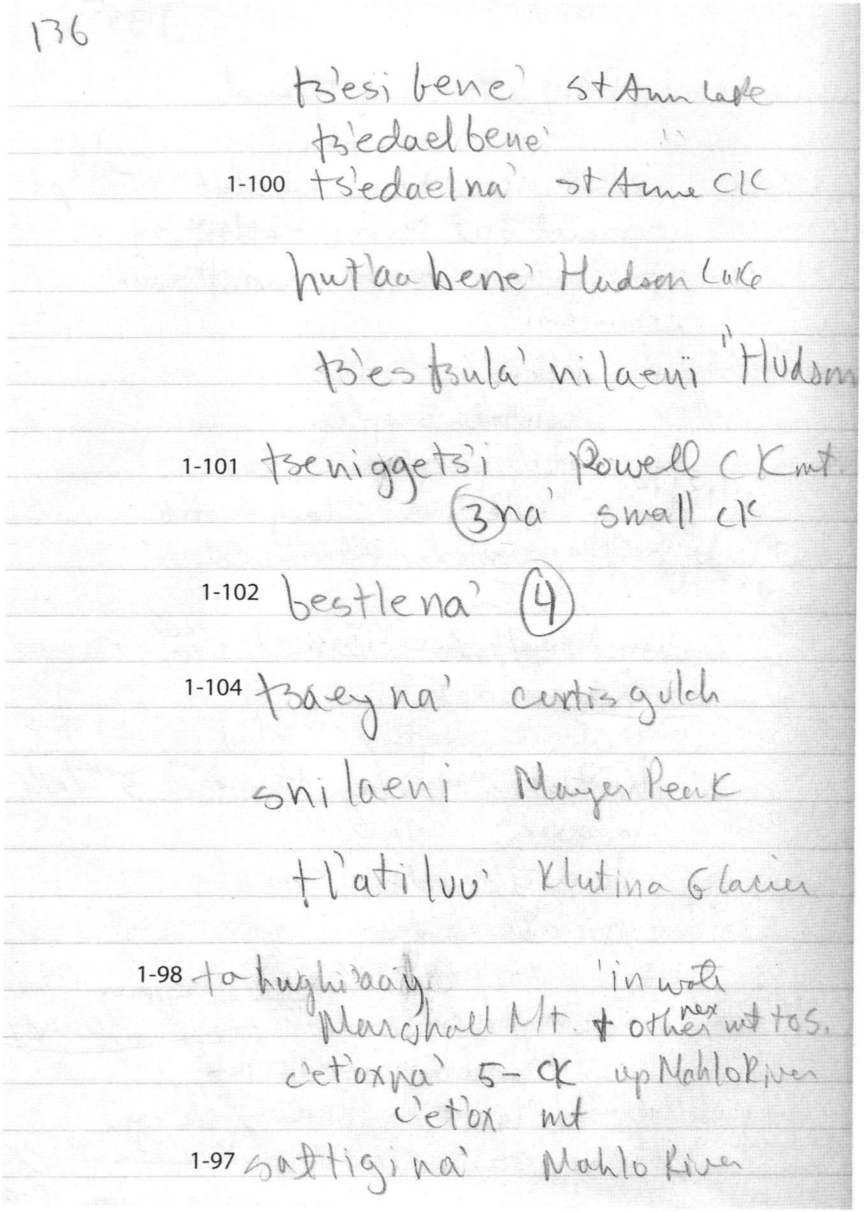

Figure 1-7. Notes from field book. Page from Kari's 1976 field notes with Jim McKinley. Fifteen place names are listed, six of which were also mentioned by Jim in his 1981 narrative.

When working with the experts on Ahtna geography every mention of a specific place name can be important. I had two sessions with Jim McKinley that were seminal for the depiction of the Ahtna place names and trail systems. The first was on April 18, 1976, and second was on January 12, 1981 (when this narrative was recorded).

Displaying his prodigious knowledge of the geography between Klutina Lake and the Tiekle River in 1976, Jim mentioned 60 Ahtna place names (Kari notebook #5:135–140). This session was not tape-recorded. When we compare these two sessions, we see that McKinley is mentally recounting overlapping geographic areas but in very different geographical sequences. In the two sessions he mentioned 84 distinct Ahtna place names. He overlapped and mentioned 20 of the same places in each of the sessions. During the 1976 session, Jim began by mentioning four distant Ahtna names, one that is on the Denali Highway and three that are on the lower Copper River in the Chitina area. Then Jim listed names in the Klutina, Tonsina, and Tiekle drainages in seven subdistricts.

Several of the names from 1976 were never obtained again from Jim nor from anyone else. Thus the repeated listings of sequences of names by experts is an important technique in copious place names research.

See, Ghanaay Zelghaende, that means udzih zelghaen, caribou.
/See, 'where a caribou was killed', that means a caribou was killed there.

I don't know how come. I guess early day maybe miracle.[16]

C'a su xa'a ts'ule'. I don't know.
/Perhaps that is so.

Caribou zełghaen, yii gha' hwdi'aan, caribou. Ghanaay Zelghaende.
/He killed a caribou, due to that it is called, caribou. 'Caribou was killed'.
3:18
Yet kangga'a gaa łdu' dae' Skesne Kuztsesden dae'.
/There to the upland here that way is 'where the Alutiiq made a mark' [1-84, bluff on S bank at 15 mi.] is thus.

Skesne Kuztsesden dae' hwdi'aan.
/'Where the Alutiiq made a mark', thus it is called.

Yet łdu' Skesne he go after us all the time you know.
/There then the Alutiiq (from Prince William Sound) would come after us all the time.

He [a war party from Prince William Sound] coming down, he coming down Klutina River. And he drown there.

Maybe 50 dae' kenii, 50 people drown there.

Yii gha' hwdi'aan.
/Thus it is named.

See, one man, one man alive there. He go up there.

That little bank, little cliff there. He go up there.

Before he go up there he painted rock.

Yii gha' hwdi'aan. Skesne Kuztsesden, Skesne Kuztsesden dae' hwdi'aan, you know.
/So that is why it is named.[17] 'Where the Alutiiq made a mark', 'where the Alutiiq made a mark' thus it is named.
4:03
Many people drown, that man, that man łdu' 'unse.
/ _____   then out in midriver.

---

[16] Jim wonders why an otherwise ordinary event should have become a place name.

[17] See discussion in Reckord 1983:141.

He go back his country, he tell that story, you know.

Yii ts'en yii ts'en we know Skesne Kuztsesden.
/From that time, from then we know about 'where the Alutiiq made a mark'.

Yet today you can see too.

Da'sdii cu gha 'skuztses k'et kut'ae.
/Still now too by there it appears like that someone has made a mark.

Yet kanggat ya'a xona.
/There the next place to the upland then.

Tsa' Bene' Gha dae' yet łnii su tsa' kuł'aani gha tsa', tsa' kuł'aani, Tsa' Bene' Gha.
/'By beaver lake' thus there he says, beaver occupy the area, beaver occupy the area 'by beaver lake' [1-85, beaver pond near Klutina R at 15.5 mi.].
4:34
Yet kangga' łdu' hwnax, Hwnax Nakey'tnelghelden.
/There to the upland, house, 'where house logs go around' [1-86, sites of miners' cabins, at 17.5 mi. or at 19 mi.].

Yi łdu' 'adii da whiteman neketezdae'łden yet hwnax neketnelghel xunt'ae you know.
/There then now the white men started to go back and forth [in 1898], there a house (of logs) was joined together there.

Yii first time they see Indian, you know.
/That was the first time they saw that.

Whiteman make it, yii yene dyilaak you know.
/They made it [cabins] that way.

Yii gha' su Hwnax Nekey'tnelghel den dae' hwdi'aan yet kangga.
/That is why 'where house logs go around' thus it is so named there upland.

Kangga' łdu' bes, bes kuts'ene hwnadadezelna'.
/The next place upland then bank, that bank (name) from there I forgot it.
(Pause) 5:07
One place two places I miss.[18]

Katsiit Hwggande Bes Ce'e yii yii c'a hwdi'aan su Hwggande Bes Ce'e.
/To the downland is 'upland big bank', it is named 'upland big bank' [1-87, hill, 2787' N of Klutina R at 16–17 mi.].

---

[18] Jim notes that he omitted this name from the Klutina River sequence.

Bes ce'e yi gha . . . yit'aa takuz'aan.
/A big river terrace is beneath that it extends to water.

Yet kanaa yełdu' yełdu' bes Hwggandi Bes Ce'e dae' konii.
/There across the way then is the bank 'upland big bank'.

Hwtsiinde Bes Ce'e hwnaade.
/Downland 'big bank' is across on other side [1-88, not in Kari 2008, bluff on S bank at 6 mi.].

That Hwggande Bes Ce'e that's big, big bes ce'e yii su taghił'aa xunt'ae.
/That 'upland big bank' is large and has water coming from beneath it, it is so.

Yii su 'adetni. Bes Ce'e, Ben tah . . . Bes Ta Tuu 'Udgheł'aayi, yii ta su 'adetnii.
/So it is said, at big bank, lake (f.s.) 'within the bank water flows past' there it is called that way [1-91, spring at bluff, at 16–17 mi.].

Two place there, you know.
5:44
Yii kanggat ya'a Bes T'aa Tu'uztaande yet su yii gha' hwdi'aan. Bes T'aa Tu'uztaande.
/There next to the upland then is 'where trail passes beneath the bank' [1-90, 'Bear Track Slide' at 20 mi., on lower terrace], that is why it is so named, 'where trail passes beneath the bank'.

Gha yet 'uyae' Klutina' hwts'en about 20 miles gaa du' kulaen I guess.
/There then that way from (here) Klutina it is about from here about 20 miles, I guess.

Yet yehwts'en xu' udi'aani sunt'aey.
/From there it is named in this way.

Yet yet kangga'a xona Dek'etl'aane Neketał'aade, that's means tuu unekeł'aa, you know.
/There, there the next place in the upland is 'current flows around cut-bank' which means that the current flows around it.
6:18
Dek'etl'aane Neketał'aade see, that's tuu 'unggu nekeł'aa xunt'ae. Yii gha', yii gha' hwdi'aan de su.
/'Current flows around cut-bank', the upland water flows around, it is so. That is why it is so named [1-91, bend on Klutina R at 21.5 mi. with steep cut bank on S side].

That's Deyitaliił de dae' hwdi'aan. Deyiitaghilen den dae' koniidze'.
/That is 'where it is flowing inside' [1-92, the Gorge area on Klutina R, 18–19 mi.], thus it is called. Thus it is said the water current flowed on inside.

Deyiitaliił.
/Current is flowing inside.

C'a yet xona yet, xungge' łdu' ben k'e dełtaan. Just deyii taghilen xunt'aedze'.
/Also there then to the upland then a lake is there. And inside that the water current is flowing inside in that way.

Naghił'aa xunt'ae yii gha' su deyii, Deyiitaliiłden dae' hwdi'aan.
/The current comes downward and due to that, inside, 'where it is flowing inside' is named.

Deyiitaliiłden yii yet kangga' c'a about three mile I think.
/'Where it is flowing inside' there then the next place upland, another three miles, I think.

C'abeni gha dae' kenii. C'abeni gha.
/'By lake that is off (the river)' [1-93, large eddy on Klutina R above Gorge at 20.5 mi.], thus they say, 'by lake that is off (the river)'.

That river that lake, river I mean.
6:51
There two places, C'abeni gha we mention, on the lake.[19]
/There are two place with that name 'by lake that is off (the river)'.

Yii same place [name] cu Ggesge Cae'e dae' hwdi'aan.
/There is the same named place also at 'skewer mouth' thus is named.

Ggesge Cae'e, you see right there Ggesge Cae'e.
/'Skewer mouth' [1-94, mouth of Manker Ck] is there, 'skewer mouth'.

That means Ggesge Na', yii gha' 'unse' k'eghey'ghiłt'esi cu gges, gges cu yii gha' hwdi'aan.
/That means 'skewer creek' [1-95, Manker Ck], due to that by the fire he was roasting something and for the skewer, roasting stick the place is named.

Ggesge Na' Ggesge Na' dae' hwdi'aan.
/'Skewer creek', 'skewer creek' thus is named.
7:21
Yet kangga yet hwts'en c'a xona 'unggat, K'eseh.
/There the next place upland from there then, a distance upland is 'outlet' [1-96, Klutina L outlet].

K'eseh dae' hwdi'aan, that's the last, end of the lake, K'eseh. End of the lake, dae' hwdi'aan.
/'Outlet' thus is named. That is end of the lake, 'outlet' is thus named.

Yet kanggadze' xona k'adii, see new place I got. I got my daddy's place su koosniic, nakuusniic gha yet.
/There from the next place upland, now I obtained my father's land, I got it back by there (Jim McKinley's homestead area).

---

[19] Jim implies that another pond is called **C'a Beni** near Manker Creek, but he does not mention this again.

I got homestead there. 160 acres I got.

Ghat yet, yet, there's another creek there, another creek.
/By there, there.
8:01
Cu yet ts'inił'aa you know 'adii desniide, Sałtigi Na' see. Sałtigi Na' dae' hwdi'aan you know.
/Another one flows out there, now I say it '? sun bump creek'[1-97, Mahlo R]. 'Sun bump creek' it is named, you know.

Yełdu' k'adii dae' yii gha' ts'utsae yedi c'a ghile', xoł c'a something,
/Then now that way the reason may be that anciently something was there, perhaps for a snare barricade.

Yedi c'a kiikedghiłt'e'i yii gha' hwdi'aani su Sałtigi Na' see.
/Something that they were using, is the reason it is so named, 'sun bump creek'.

Sałtigi Na' dae' hwdi'aan, that's little ways there.
/'Sun bump creek' thus it is named.
8:26
K'adii łdu' white man they call Salmon Creek, dae' ł'aan, dae' kuyighize', I don't know. That right c'a su.
/Now then white people call that one 'Salmon Creek,' thus they named it. I don't know. It seems that's right.

Yihwts'en kanaa 'unggat łdu' Tahwghi'aay dae' konii
/From there directly across and to upland then is 'one that extends to the water' thus is said.

Tahwghi'aayi nic'akuni'aade, xona Tahwghi'aay dae' konii.
/'One that extends to the water' [1-98, Marshall Mt] is an area that extends out from the shore, then 'one that extends to the water' thus it is said.

Tahwghi'aay.
/'One that extends to the water'.
break 8:53
That Lucy Brenwick k'adii ye zdaa, that place there.
/That Lucy Brenwick now she stays at that place there.

That's only place su Ts'abael Nic'ani'aay gha konii. Ts'abael Nic'ani'aa you know.
/Only there by 'spruce extends from shore' [1-99, on N shore Klutina L at Lucy Brenwick's homestead] it is said. 'Spruce extends from shore'.

Yii gha' c'a su ts'abael we always call Ts'abael Nic'ani'aade that's better.
/That is why, spruce, we always call it 'where spruce extends from shore', that's better.

That's really call[ed] that way.

Ts'abael Nic'ani'aade.
/'Where spruce extends from shore'.

That's what we gonna call.

Yet kanaa 'unggat yełdu' Tahwghi'aay, Tahwghi'aayi dae' mountain 'utsene.
/There directly across and to upland then is 'one that extends to the water' [1-100], 'one that extends to the water' is to the lowland (toward the lake).

Tuu, ben yii taghi'aayi gha su, mountain ben yii taghi'aayi yii gha' su Tahwghi'aayi dae' hwdi'aan.
/To the water, into the lake it (ridge) comes to the water there, it extends to the water is the reason it is named 'one that extends to the water.'

Tahwghi'aayi.[20]
/'One that extends to the water'.
9:29
Du' yet kanaa xu ya'a Ts'edael Ts'edael Na', you see dan'edze' Ts'edael Na'.
/Then there across the way is Ts'edael, 'Ts'edael creek' [1-100, St. Ann Ck], from the upstream is Ts'edael Creek.

Yet kangga' k'a Tsinigge'si Tsinigge'si dae' koniiyi.
/Right there next to the upland is 'twisted head', thus is said 'twisted head'.

Tsinigge'si yii łdu' mountain, yii gha' hwdi'aani su Tsinigge's yet 'unggat.
/'Twisted head' is that mountain, thus is the reason it is named 'twisted head' [1-101, West Peak, Powell Peak] to the upland.
Du' yet kangga' xona Bestle Na', Bestle Na' dae' hwdi'aan.
/Then there the next place upland is 'little bank stream' [1-102, Curtis Gulch], thus is named 'little bank stream'.

That's bes ts'e', bes uyidighił'aayi c'a.
/There toward is a river terrace, the current flows into the terrace.

Bestle Na' dae' hwdi'aan c'a su.
/Thus it is named 'little bank stream, it seems.

Bestle Na', Bestle Na' dae' hwdi'aan see.
/'Little bank stream', 'little bank stream' it is named, see.
10:05
Dii yet kangga'a xona Łuk'ae Tu' Na' dae' hwdi'aan. Łuk'ae Tu' Na'.
/The next place to the upland then is named thus 'fish soup creek' [1-103, Hallet R]. 'Fish soup creek'.

───────────

[20] See Figure 2-2, on the duplication of this name on both Klutina Lake and Tazlina Lake.

Łuk'e Tu' Cae'e dae' hwdi'aan. that's something łuk'ae kez.
/'Fish soup mouth' [1-104, mouth of Hallet R] thus is named.

Yenida'a tah yet c'a łuk'ae kezdlaets, you know something, yii tu' yii gha' c'a su hwdi'aan. Łuk'ae Tu' Cae'e.
/In ancient time there they boiled some salmon, or something and due to that, that soup, is why it is so named. 'Fish soup mouth'.
10:29
Yet kanggat biyudghiłeni yii su' Tsaey Na' dae' hwdi'aan.
/There the next place upland, the current that flows past there, that one is named 'tea creek' [1-105, creek from Terrace Mt].

Tsaey Na' yii c'a 'adii that's easy we know that.
/'Tea creek' too now.

Tsaey Na' tsaey, dae' people tsaey nadelna'.
/'Tea creek', tea, people forgot their tea on that creek.

They forget tea up there on that creek, you know.

Yii gha' hwdi'aan su, Tsaey Na'
/That is why it is named 'tea creek'.

Tsaey Na', Tsaey Na', tsaey people forget tea. Yii gha' hwdi'aani su Tsaey Na'.
/'Tea creek', 'tea creek', people forgot tea. That is why it is named 'tea creek.'
10:57
Yet kanaat ya'a yi ben diłende yet kanaat k'a Tay'laxden dae' hwdi'aan.
/There the next place across there, right there where current flows into the lake, there next place across is 'spawning in water place' [1-106, creek into Hallet River from S] thus it is named.

Tay'laxden that's fish go up there. That place there, yii gha' hwdi'aan.
/'Spawning in water place', fish run there. That is why it is named this.

Tay'laxden.
/'Spawning in water place'.

Yet kangga'a xona Bes Nilaeni dae' hwdi'aan, Bes Nilaeni, that's mountain.
/There the next place to the upland 'river terrace one' is so named, 'river terrace one'.

That's big mountain, debae yet ik'etił'as you know. Bes Nilaeni dae' hwdi'aan.
/Sheep move about upon it, 'river terrace one' [1-107, Meyer Peak] so it is called.

Yet kanaa ya' k'a xona Ce's Di'aedzi Cae'e, Ce's Di'aedzi Cae'e
/There the next place across right there then is '? spear stepping mouth', '? spear stepping mouth' [1-108, mouth of creek into upper Klutina R].

Figure 1-8. 1898 photo of upper Klutina Lake viewed toward the northwest from the mouth of Hallet River [1-104, Łuk'ae Tu' Na'].

Photo by Frank Schrader,
U.S. Geological Survey,
in 1898. Courtesy of Bill Simeone,
Alaska Department of Fish and Game.

That's a something miracle [i.e., unusual about this] name too, I don't know.

Ce's Di'aedzi Cae'e yet su cu we don't know hw'eł ts'esniige.
/'? spear stepping mouth' is there, but we don't know the meaning.

C'a hwgha hwdi'aandze' hw'eł ts'estniige, you know.
/I don't know why it is so named.

Dae' Ce's Di'aedzi Cae'e.
/Thus is '? spear stepping mouth'.

Maybe that's yenida'a c'a su.
/Maybe it is from the ancient times.

Maybe saghaniggaay yii gha' c'a su hwdi'aan, I don't know.
/Maybe by Raven for some reason it is so named.
11:59
That's I don't know that. I just can't mention that one.

Gaa yet Ce's Di'aedzi Cae'e yet kanggat c'a xona,
/Here there at '? spear stepping mouth' the next place to the upland then,

Tsiyidełtaani dae' hwdi'aan.
/'The one (pond) that is within rock' [1-109, two lakes N of Ce's Di'aedzi], thus it is named.

Tsiyidełtaani, that means cliff go around that lake, that mountain.
/'the one (pond) that is within rock'.

Little mountain cliff go around there and lake is inside there.

You see yii gha' hwdi'aani su. Tsiyidełtaani, lots fish go up there.
/You see that is why it is so named. 'The one (pond) that is within rock'.

There's two lakes there, nadaeggi two lake there, there's lots fish going up there.
/   two

Yet katggat yak'a xona Hwninighinaesi. That's the mountain, Hwninighinaesi see.
/Right above there then is 'long one comes to a place', that mountain is 'long one comes to a place' [1-110, Black Mt and ridge 5767'
'Murray' on East Fork of upper Klutina R].

Big long mountain there, cliff. Cliff mountain big.

Yii gha' hwdi'aan, that Hwninighinaesi means.
/That is why it is named, 'long one comes to a place'.

All the way down to Ce's Di'aedzi Cae'e that's only one mountain there they call Ce's Di'aedzi,
/'? spear stepping mouth' [1-110] '? spear stepping',

Ce's Di'aedzi Dghilaaye' that's what they, I mean Hwninighinaesi Dghilaaye' they call.[21]
'? Spear stepping mountain', I mean 'long one comes to a place mountain'.
13:00
Gaa yet c'a kanggat yak'a xona łic'udghidlenden yet, Łic'udghidlenden.
/Here the next place upland right there is 'currents join together place', 'currents join together place' [1-111, confluence of branches
of upper Klutina R].

That's means three water come together, you know. Come together. One is north, one is south, one is Valdez Glacier. Valdez Glacier
water flow through together, see.

Yet kangga'a xona bes Łuu k'a Łuu T'aa dae' hwdi'aan, Łuu t'aa.
/There the next place upland, is a terrace at the glacier, 'beneath the glacier' [1-114, base of Klutina Glacier'] it is named, 'beneath the
glacier'.

Łuu 'adii ts'utsae whiteman yet 'u'ane uk'eteni te'eł kughile'.
/At the glacier anciently there before white men a trail upon it went over to the other side.

---

[21] Jim corrects himself.

Whiteman, whiteman tes ik'e łunidae'ł you know.
/The white men came over it (pass) on it (the trail).

That łuu ubaaghe Valdez nse'.
/The edge of that glacier (goes) out to Valdez [1-115, Valdez].
13:44
Yii gha' hwdi'aan Łuu T'aa dae' hwdi'aan, Łuu T'aa.
/That is why it is named 'beneath the glacier', 'beneath the glacier' [1-118].

Yii I go up there Nałudelc'edi dae' koniiyi,
/There I went up there too, it is named 'glacier that stretches across' [1-116, summit between Klutina and Valdez glaciers].

That's right on top that mountain, on top that glacier there.

Yii yii su Nałudelc'edi dae' hwdi'aan.
/There it is named 'glacier that stretches across'.

On top go up through before you go through pass to Valdez, you know.
14:06
'Utggat Nałudelc'edi yihwts'en 'utsene 'adii cu łu su close 'ughezdlet you know.
/Up above 'glacier that stretches across' from there to the lowland it has melted close by.

'Ughezdlet xu' dyaak you know.
/It has been melting, you know.

There's no glacier now, you know. Before you see it way clear down to the creek, Valdez side, you know right clear down to the creek. Glacier go down that far, see.

Now is nothing. You go pretty near clear up to the top you never see glacier no more. That's all thawed out there. Somewhere you know. That's all top of the glaciers you see going down there about three, four hours, top of the glacier. Go down to Valdez. Four hours.

'Utggah ts'en top Nałudelc'edi gha hwts'en 'utsiit Valdez hwts'e'.
/From the top at 'glacier that stretches across', to the lowland at Valdez.

About four hours kulaeni k'et kut'ae, you know.
/It is about four hours, you know.

I walk down that time, see. That time was glacier, now it's, now it's no glacier there no more now, see.
15:08

Whiteman hwteshghidae'ł kiiłnii.
/The white people went over the pass there.

You see one glacier, utes yighi'aax xunt'ae you know. This side naghi'aax xunt'ae.
/The glacier goes up over the pass there. It (glacier) extends downward on this side.

Whiteman, early day, 1898dze', whiteman tes ghidae'ł kenii xu xu su.
/The white man in 1898 came over the pass there, they say.

K'adii gha hne'sdelghosde.
/Now we are talking about that.

'Utsii yet he got right foot of the glacier it's about about fifteen graveyards there k'adii you know. Fifteen graveyards there see.
/In the downland   now
15:37
U . . . ghu 'utggu łuu c'u yet yighidax nt'iix.
/*f.s.* Just above where the glacier is coming in.

Ye 'utggu 'utsii danakuhdaghalyaes ya'a.
/Above there and in the downland they buried them, right there.

He's [they're] buried up there. There you can see that fifteen graveyards there today.

You know cu yet xunggu xu'ane su kakudaan k'et kut'ae, you know.
/Also there in the upland on the other side is a cavernous, tunnel-like valley.

Dae' Hwninighinaes k'adii de. Dae' Hwninighinaesi.
/That is 'long one that is embeddded'. Thus 'long one that is embedded' [1-110, Black Mt and ridge 5767' 'Murray' on East Fork of upper Klutina R].

Yii ghanggu yae' kakudaan tkut'ae.
/There to the upland of it (that valley) is cavernous.
16:00
Yii kanggu su 'un'e nani'aax xunt'ae you know.
/There in the uplands upstream that (mountain) extends across.

Yet yet yii su dadaa' ts'en Tonsina' hwts'ene, 'usghu, Little Little Kentsii na' dae' kenii.
/There, there to the downstream side is the Tonsina River side, out in front there is Little Tonsina River, thus they say.

Kentsii Na' Ggaay dae' kenii.
/'Little spruce bark boat stream', thus they say [1-115, upper fork of Greyling Ck, S of mountain 1-110].

Yii Kentsii Na' yidighiłeni su.
/It flows on into 'spruce bark boat stream' [1-116, Tonsina R].

Yii t'aa, yehwdi'aani su Basnats'ests'elyaesi see,
/Beneath that is the one named 'someone put rocks on the fire by it' [1-117, Black Mt 6198' W of Greyling Ck],

Basnats'ests'elyaesi.
/'someone put rocks on the fire by it'.
16:29
See means ts'utsaet, you know, 'utgga, 'utggat debae ts'eghił'as yełdu' c'a su.
/You see that means that long ago, you know, up above, up above sheep would come out it seems.

'Utsii yet camp c'ekot'iix, you know.
/In the downland there they customarily had a camp.

Yihwts'en 'utggax kiineł'iix, you know.
/From there they would look above, you know.

'Utgga łuu 'utggat ts'ił'as you know.
/Above, above the glacier they (sheep) would come out.

'Unggat ye xał'udasdenta nanest'aan xunt'ae dze'.
/In the upland there that (mountain) seems to be standing out by itself.
16:41
Yełdu' nggat ngga yet kidaltlet dze' ałdu' gha'uk . . . danggedze' nahyuniyiis.
/Then in the upland, in the upland there, *f.s.,* they would leap at them (in pursuit) and then they would climb down from the uplands.

Ghangga gha yet ku'ak'eh nayel'as yełdu', all they kill 'em, you know.
/And upland of there they (sheep) would move in between them and they would kill them.

Yii gha' hwdi'aani Basnats'ests'elyaesi.
/That is the reason it is named 'someone put rocks on the fire by it'.

Yehwts'e dadelts'edi yet su ts'utsae you know.
/From that place there these words have come there long ago.
16:58
Ts'utsaet 'adii 'talaał c'a su talaał' dae' hdghine' koniide you see.
/Long ago now 'there will be many, there will be many', thus they said it would be.

'Utgga gha yet yi hwts'idalts'edi su.
/Up there there these words have come from there.

Natggat łu k'adii dana'idyaax xu.
/Up above now he (hunter) enter (the sheep herd) again.

'Utggat ts'inac'il'aa's you know, tuu nac'il'aa's n'eł de
/They (sheep) would moved back out again from above, and they would go to the water and

'Natggadze' nac'il'aa's natgga' cu natggatde.'[22]
/'They have moved from above, up above, again up above.'

He make three, two trips he make already, see. And then third time, see, then he gonna go, you know. And then he say,
17:24
'Talaał xu su talaał.'
/'There will be many, there will be many.'

Yehwts'en dadelts'edi su 'adetnii.
/These words come from that time, it is said.

'Talaał xu su talaał,' nii.
/There will be many, there will be many,' he says.

He dress 'em up and he go up again. That's three time he make that trip that day. Same day.

Yii gha' hwdi'aani su, yii gha' su Basnats'ests'elyaesi dae' konii.
/That is why it (the mountain) is so named, thus it is said, 'Someone put rocks on the fire by it'.
Ends 17:47

---

[22] Apparently this is a statement by someone observing the hunter from below.

The Klutina River Drainage Through Time

Jim McKinley's account is the only first-person Ahtna narrative of travel to Valdez via the Klutina Glacier. Sometime in the early 1920s McKinley made this trip from Klutina Lake to Valdez with his father, McKinley George. For two to three years, 1897 to 1899, about 3,000 transients—men traveling from Seattle through the new town of Valdez and lured by the Klondike Gold Rush—used what was an ancient Ahtna trail.

Two statements about this trail are on record from the first two U.S. Army expeditions up the Copper River, led by William Abercrombie in 1884 and Henry Allen in 1885. In the summer of 1884 Lieutenant Abercrombie led a well-equipped Army expedition of 15 men to Copper River and was charged to "ascertain as far as practicable the numbers, character, and disposition of all Indians living in that section of country" (Abercrombie 1900:383). As Abercrombie returned to Prince William Sound, his group traveled toward Port Valdez, and on September 11, 1884, in the vicinity of Ellamar or Tatitlik he met an elder Russian Aleut, who informed him:

> Some years ago, probably—twenty or thirty, the portage was used entirely by the upper River Indians, who came down Copper River to a stream heading in the lake, which not being previously named or visited by white men, is designated as Lake Margaret [Klutina Lake]. Up this they traveled to the lake, hence to the foot of the passage. Here they left their bidarra and packed their furs over to salt water, where bidarras were furnished by the Chugachmiutes for the voyage to Port Etches. For this service the Upper River Indians paid a tribute in furs, not only for the bidarra, but also for the privilege of passing through the country.
>
> On one of these periodical tours there arose some disagreement, and words led to blows. The result was the annihilation of the Copper River party by the Chugachmiutes.This occurred in the spring. In the fall an epidemic of some kind carried off over half the inhabitants of the village, which caused the remainder to flee not only from the epidemic (which they attributed to the powers of the shaman of the Copper River natives), but from their wrath. I believe these people should be feared. After the flight above referred to, the Upper River Indians adopted the route via the mouth of Copper River. It is now used by them in the month of March, when they come down on the frozen river, in May after the break-up, and in September when the river has subsided. (Abercrombie 1900:383–84)

The next year, 1885, the three-member Allen expedition passed the mouth of the Klutina River on May 13. Allen recorded the name "Klutena," the Ahtna name Tl'atina'. Allen (1887:62) confirmed what Abercrombie had heard about the Ahtna trail over the glacier to Prince William Sound the previous year:

Figure 1-9. Jim McKinley and his wife, Elephina, in the late 1930s with their children, Tom McKinley and Cecelia McKinley (Larson).

Photo courtesy of Ahtna Heritage Foundation.

Midway between the camps we passed the mouth of the Klatená, the largest tributary of the Copper save the Chittyna yet passed, and a stream of size, as shown by the general topography of the country. The mountains on the west side, as far as the eye could reach, seemed to be separated by this river. Natives report this river to head near Nuchek in a large lake, where fish are abundant; that to reach Nuchek, however, would necessitate the crossing of large glaciers. Nicolai informed me that he had been to its source when he was a small boy. In accordance with his recollection I have traced it in dotted lines on the accompanying map.

Since Chief Nicolai was elderly in 1885, this implies that the Klutina River/Lake trail over Valdez Glacier to Valdez Arm was regularly used until the 1850s.

In 1884 Abercrombie recorded this information about another Ahtna trail between Copper River and Valdez Arm: "I had been informed by the upper river Indians and those on the coast that a shorter route existed via Port Valdez over the mountain to the lake, the outlet of which ran into Copper River below the Chyttyna" (Abercrombie 1900:389). The route Abercrombie was told about is likely the Tasnuna River–Marshall Pass–Lowe River trail to Valdez Arm. In 1885 this route was reconfirmed and mapped by Allen. See Figure 0-4. It is noteworthy that the two main routes into Copper River (Klutina Glacier pass and Lowes River connecting to Thompson Pass) in the gold rush period (1898–1902) were trails that the Ahtna used and that they were documented by Abercrombie and Allen in 1884 and 1885.

Jim McKinley's lifelong use of the Klutina River and Klutina Lake areas was well known. He is mentioned numerous times in the recent Klutina River navigability study (Brown 2004:9, 23, 28–29, 31, 32, 33). One incident (Brown 2004:30–31) is especially dramatic:

In 1921 Shirley A. Baker of the Alaska Fisheries Service visited

Klutina Lake to check on the health of the salmon run. . . . Baker's report is of particular interest because he described a meeting with several Indians and the Indians' success in descending the Lower Klutina River in a boat. While Baker and McCrary were camped at the lake outlet, Chief McKinley Jim, Big Charlie, and four other Indian hunters crossed the lake to visit them. . . . Baker and McCrary left the lake on October 12 and reached Copper Center the following day. On the way down, they kept a close lookout for Big Charlie and two other Indians who had left Klutina Lake the day before in a skin boat. Baker's account of the event is worth quoting in full:

> Big Charlie's wife was very ill from an attack of inflammatory rheumatism, and it was impossible for her to walk or ride horseback, so Big Charlie and the chief held a pow wow and consulted with the guide and me about the feasibility of coming down the Klutina River in a staunchly made skin boat which they had just finished constructing out of goat and sheep skins in order that they might take the almost helpless woman to a physician. We told the Indians that such a journey was feasible all right, but that it would be necessary to have a good man to steer from the stern of the boat and another good man to keep the bow of the boat headed upstream and row like the devil all the time,—especially when approaching the numerous treacherous places where there arc large boulders. We advised the Indians to line the boat downstream when they reached Devil's Elbow and Horse Shoe Bend, two of the most dangerous spots in the river. Noticing so many swift, rocky places that we had not observed before while we were going down the trail on the bank of the river, we became somewhat uneasy about the Indians, wondering whether they had succeeded in passing through these ugly waters safely.
>
> Immediately after arriving at Copper Center that evening, I called to see the Indian chief and told him about the party coming down Klutina River with the sick woman. As they had not been seen or heard of since leaving the lake, I suggested that a search party be sent out. The Indians arranged to do this the following morning.
>
> October 14: At Copper Center. As the search party was ready to start, we heard some peculiar Indian yells, and looking up toward the Klutina River bridge saw Big Charlie rowing, for all he was worth, the little skin boat which held his sick wife and the other Indian who did the steering. They were dripping wet from the splashing waters, but they were very happy and proud of the fact that they had safely navigated the swift Klutina River.[1]

Also from Brown (2004:25):

From time to time, the local newspapers carried stories that mentioned travel on the Klutina Lake trail. In the spring of 1939, for example, newspapers reported that Jim McKinley of Copper Center, whose wife had died that winter, took his family of four children to the lake for beaver trapping season. He returned to Copper Center in early June.

---

[1] Shirley A, Baker. Assistant Agent, Alaska Fisheries Service, to Commissioner of Fisheries, November 10, 1921; and Shirley A. Baker, "Chronological Report Covering Investigations on the Copper River, Oct. 1 to Oct. 18, incl., 1921," Copper River, 1921, Reports of Bureau of Fisheries Agents, 1917–29, Central Classified Files, RG 22, NA. Microfiche at Alaska Resources Library.

2

Ben K'atggeh hwts'e' Denaey 'Iine 'Uze'

Personal Names of Chiefs Toward Tyone Lake

Figure 2-1. Frank Stickwan speaking at the Ahtna Culture Camp in 2001.

Photo courtesy of
Ahtna Heritage Foundation.

Frank Stickwan

The home territory of Frank Stickwan (1902–2005) was Dry Creek Village on the Copper River and the area south of Crosswind Lake. He was intimately familiar with other drainage systems that feed into Crosswind Lake from the north, including the West Fork and Middle Fork of the Gulkana River. His career travel map extends north to the Delta River and Tangle Lakes areas. He also traveled to the upper Dadina River on a hunting trip in 1923. For many years Frank and his wife, Elsie, trapped at Game Trail and Salmonberry lakes, where they had cabins, and on the west side of the Copper River (Klawasi Creek area).

In our fisheries interviews of 2000–2002 we were highly impressed by the vast factual knowledge Frank had of the hydrology, fisheries, and climate of the general area between the Gulkana and Tazlina rivers. Like Jim McKinley, Adam Sanford, Fred John, and other Ahtna experts, Frank has a very broad secondhand knowledge of other areas through the Ahtna oral traditions.

Between 1975 and 2003 I worked with Frank more than 10 times. Six untranscribed recordings with Frank focus on resources and travel and **yenida'a** stories. This short segment is from a nearly two-hour recording session on October 2, 2000, in Tazlina. This is a good example of the Ahtna association of the personal names of famous chiefs with important sites and settlements. Ahtna personal names play a very important role in Ahtna culture.

Starting point: Tazlina village
Ending point: Tyone Village
Total is 6:40, from AT107 (5:00 to 12:00). Sound file chp2-frankstickwan.wav.

That's all. Xona yehwk'e yaen' de nahwnic łdahwdghelnen.
/That's all. So only to here is the information of great antiquity.

Dae' Tezdlen Na' xutgga' hdaghalts'e' ne yene de.
/That 'swift current stream' [2-1, Tazlina R], up above there people were staying.

You want to know too, huh?

Koht'aene k'et?
/In the Ahtna language?

JK: Yeah.

Yene c'a hw'eł kutnes.
/You ought to know about them as well.

Tazlina Lake and Tyone Lake hdaghalts'e'ne.
/Those people who were staying at Tazlina Lake [2-2] and Tyone Lake [2-3].

Tezdlen Na' 'utggat Taltsogh Na' hwdi'aan den Taltsogh Na' 'unggat.
/'Swift current stream' [2-1, Tazlina R], above to what is named as 'yellow water stream' [2-4, Talsona Ck], and 'yellow water stream' in the uplands.

'Utsiit Taltsogh Cae'e, yihwts'en tighita' 'unuux Nkaał Bene'.
/The lowland place is the 'mouth of yellow water' [2-5, mouth of Talsona Ck], from there was a trail to the upstream area of 'salmonberry lake' [2-6, Game Trail L].

'Adii 'adii diniiyi, you know, Seven Eggs Lake udiniiyi.
/Now, now you call that, you know, Seven Eggs Lake you call it.
0:48
Yi baaghe k'etighita'.
/The trail comes onto its shore.

'Uniit I'dzak'ehi yii baaghe 'eł.
/Upstream at 'game trail' [2-7, Salmonberry L] to its shore too.

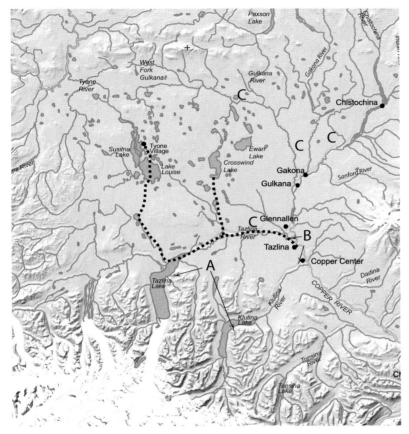

Figure 2-2. Frank Stickwan's route in Chapter 2 covers about 80 miles. Some duplication in Ahtna place names is distinctive. A, B, and C indicate instances of duplicate place names. It is certainly noticeable that on the large L-shaped glacial lakes, Klutina Lake and Tazlina Lake, two ridges at the south corner-points on both lakes have the same name, indicated by A: **Tahwghi'aayi** 'one that extends into the water.' Also the first sizable north-side bluffs on both the Klutina River and Tazlina River have the same name, indicated by B: **Ts'inahwnet'aaden** 'object that extends straight out'. Ahtna experts are well aware that four streams have the same name, **Taltsogh Na'** 'yellow water stream'. C on this map indicates the four duplicate names: Talsona Creek [2-4], Sourdough Creek, an officially unnamed creek above the West Fork mouth, and Tulsona Creek [1-57].

Kaghalk'edi baaghe n'eł.
/And to the shore of 'protrusion comes up' [2-8, Crosswind L] too.

All that trail everywhere.

Koht'aene tene kughile'i.
/The people's trails were everywhere.
1:00
Taltsogh Cae'e ghida'en yen Scołta' udghi'a'en.
/The one that stayed at 'yellow water mouth' [2-5, Talsona Ck mouth] he was named 'father of Scołta' (meaning uncertain)'.

Scołta', maybe you have it [that name] before too I guess. Scołta'.

Xona yehwts'en 'unggat Bendil . . . Bendildende 'ehwdi'aan de, yii c'a.
/Then from there to the upland is called 'lake flows place' [2-6, mouth of Mendeltna Ck], that place.

Taaden koht'aene hdaghalts'e' kanii Bendilden Bene', ubaaghe ina.
/In three places the people stayed upstream on 'lake flows lake' [2-2, Tazlina L], the ones on its shore.

Tezdlen Na' Uyidadiniłen de yi c'a ugheli yii c'a koht'aen hdaghalts'e' kenii de.
/'Where it flows into swift current stream' [2-7, site N of outlet of Tazlina L], there too is good place, there too people stayed, they say.
1:30
Yet kanggu hdaghalts'e' ne.
/People stayed in the upland there.

Ts'i'sc'elaes Ta' yet that was his name there. Tazlina Lake Uyidadiniłen den.
/'Father of we throw things out' there at 'where current flow into Tazlina Lake [river]' [2-7].

Ts'i'sc'elaes Ta', yeah.
/'Father of we throw things out'.

And yehwts'en on this side, yedu' yenaa ts'en K'aay Na' dae' hwdi'aan de.
/From there, the across the way side is called 'ridge creek' [2-8, Kaina R].

Tazlina Lake yitadighiłen.
/The current flows into Tazlina Lake.

Łuk'ece'e una' c'ilaen uyii, yi c'a.
/King salmon are in that stream.
2:00
Sii 'ele' hwneł'iile xu'nt'ae.
/I have not myself seen that place.

Yet xuk'e Kulaaghe Ta' ak'ae kughile'. Kulaaghe Ta'.
/There was the home of 'father of on their behalf.'

Kulaaghe Ta', Kulaaghe Ta'.
/'Father of on their behalf,' 'father of on their behalf'.

Yet c'a xona Bendilden, Bendilden Tazlina Lake Tazlina Bene' yitadiniłen de.
/There then too at 'lake current place' [2-6, site at mouth of Mendeltna Ck], where that (stream) flows into Tazlina Lake.

Yet łu xona łuk'ae una' tedełde, Bendilna'. Ye hwdi'aan dze' Bendilden.
/There then salmon go into that steam, 'lake current stream' [2-7, Mendeltna Ck]. It is named 'lake current place'.

Yet cu xona, yet cu xona more there c'eyits'e' tnaey daghalts'e'ne yene.
/There also then, there the most head man stayed there, they did.

Ts'e' Idahwdetnes Ta' ye ghida' in there. Ts'e' Idahwdetnes Ta'.
/'Father of there is a sound toward him' stayed there. 'Father of there is a sound toward him'.
2:36
All łuk'ae gha yet da kanahdaghalts'e' ne.
/All the people customarily stayed there for salmon.

Sc'et'exen c'a yet ghida'.
/(*Sc'et'exen*, meaning uncertain) also stayed there.

Stakolc'eł yet yighida' yet.
/'He tore it off' stayed there.

Baniłtah yen c'a yighida'.
/'Together by him' he too stayed there.

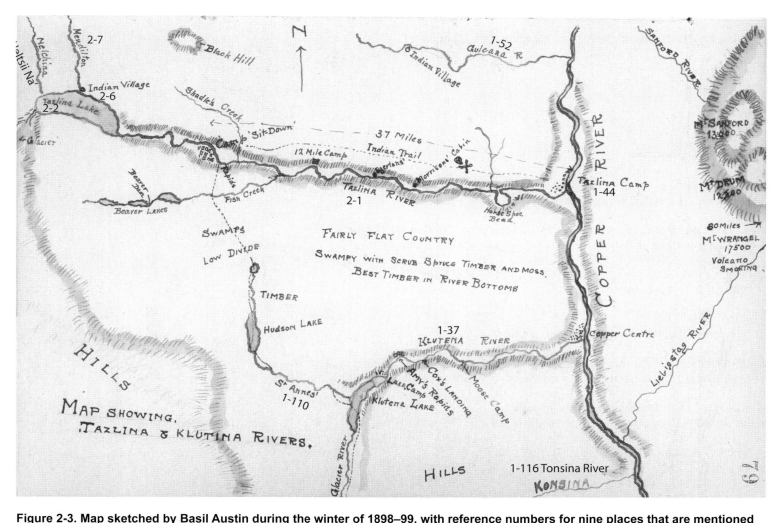

Figure 2-3. Map sketched by Basil Austin during the winter of 1898–99, with reference numbers for nine places that are mentioned in Chapters 1 and 2, as well as the name Neltsii Na' **for Little Nelchina River.** Austin stayed for a few months in a makeshift cabin on the Tazlina River at Twelve-Mile Camp. Chief Nicolai of Tazlina Lake befriended Austin and oriented him to trails and place names. Austin's map notes in some detail major Ahtna trails on the north terraces of the Tazlina and Klutina rivers. He notes the Ahtna village on lower Mendeltna Creek; Hudson Lake here is actually St. Ann Lake. The name "St. Annes" recorded here was (humorously) reshaped from the Ahtna name **Ts'edael Na'** [1-100], the meaning of which is uncertain (Austin 1968).

Bendilden yene.
/They are the ones of 'lake current place' [2-6, Mendeltna Ck mouth site].
3:00
What you call 'em,

nexunae'
/upriver of us

Nikolai Lake hwdi'aan de xu'.
/[2-9] Nikolai Lake it is called.

Yet łu Ni'sdela' Ta'.
/There it seems there was 'father of puts things there'.

Nikolai Lake, Ben Daes Bene' dae' u'di'aan, I think, koht'aene k'e. Ben Daes Bene'.[1]
/Nikolai Lake 'shallows lake' it is called, I think, in Ahtna. 'Shallows lake.'

Ts'e' desnii den Ni'sdela' Ta'.
/So I say, 'father of puts things there'.

Ni'sdela' Ta', Ni'sdela' Ta'.
/'Father of puts things there', 'father of puts things there.'
3:37
Nes . . . Nesteni Uta' cu dae' another one, another one Nesteni Uta' cu yet ghida'.
/'Father of frozen face', also is another one who stayed there also, 'father of frozen face'.

Gha ye old man. Ben Daes Bene' su ye ghida' in there.
/There he was an old man who lived there at 'shallows lake' [2-10, Old Man L].
4:14
Ben Daes Bene'. C'ec'adax, C'ec'adax, at Old Man Lake.
/At 'shallows lake' [Old Man L] was 'things drop down', 'things drop down'.

Lots of people around there too long time ago, Lake Louise ubaaghe, you know.
/   its shore. [2-11, L Louise, Sasnuu' Bene']

Nihwneldiił Ta' yen c'a yen c'a xu' ghida' you know.
/'Father of ? turning red again', he too stayed there.

---

[1] Actually, Frank here calls the name of Old Man Lake.

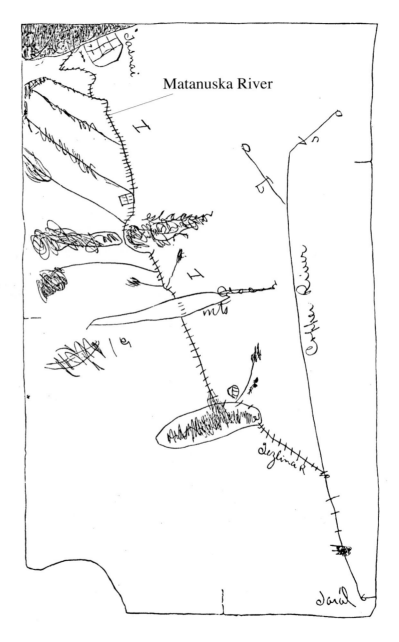

Matanuska River

Figure 2-4. Sketch map of the trail along the Tazlina River to Tazlina Lake, the Matanuska River to Knik Arm obtained by Henry T. Allen in 1885 from an Ahtna man.

The name **Tezdlen Na'** [2-1, Tazlina River], has been independently reported by Ahtna speakers in 1797, the 1830s, 1848, 1885, 1898, and many times later.

Allen (1887:61) wrote: "I had frequent maps made by the natives to show us the trail over the Alaskan Mountains and down the Tanana, to the Yukon River, all of which indicated the route to be via the Tez-lina River to Tasnai [the Ahtna name **Dastnaey** for the Dena'ina]. The accompanying sketch represents one of the maps thus made. . . . It was not until we reached the Tezlina that I felt sure the trail up it could not lead over the Alaska Mountains but rather to Cook's Inlet." Features on the sketch map: a house or village at Mendeltna Creek on Tazlina Lake, the trail and pass to the Matanuska River, several tributaries of the Matanuska River, a house on the south side of the Matanuska, and a village site near Knik Arm.

The earliest source on Ahtna language and geography is Russian navy man Dmitri Takhanov, who in the spring of 1797 wrote down about 20 Ahtna words and 12 recognizable Ahtna place names (Black 2008:78–80). While at the mouth of Tonsina River he recorded from three Ahtna men in order, eight or nine identifiable trail-route place names from the mouth of the Tonsina River, to the Tazlina River, to Tazlina Lake, to the Little Nelchina River, to the upper Matanuska River.

Lake Louise baaghe. Nihwneldiił Ta',
/On the shore of Lake Louise 'father of ? turning red again.'
pause
Lots of old people up there, I forgot all that names.
5:08
Way up Tyone Lake, Tyone Lake another U'eł Yayaał Ta' udi'aan yighida'en.
/At Tyone Lake [2-3, Ben K'atggeh'[2]] another guy was named 'father of he walks with him', he stayed there.

U'eł Yayaał Ta'.
/'Father of he walks with him'.

And ts'i . . . Ts'iidak'aał Ta', Ts'iidak'aał Ta' that's his father that man.
/ And 'father of filing it off,' 'father of filing it off'.

Ba'ane 'Sc'eyelyaas Ta', Ba'ane 'Sc'eyelyaas Ta', kudaghalts'e'ne long ago.
/'Father of someone brings things from beyond (Ahtna country)', 'father of someone brings things from beyond' they stayed there long ago.

And what you call em. And another man C'enih, C'enih.
/   (possibly) 'he says something'.
6:00
C'enih, yen c'a that same place, you know. Xuya'a, C'enih.
/'He says something' too stayed at the same place, right there, 'he says something'.

All yet Tyone Lake baaghe hdaghalts'e'
/They all stayed on the shore of Tyone Lake

uts'en ne'aayi 'udełende.
/where the current flows by on that side.
6:15
There's [were] lots of [other] people, we don't count them people, you know.

Just the highest people that's all we talk [about].

That's their own village, they take care of their village, you know.

That's the way people live around there.
Ends 6:40

---

[2] Chief Tyone, the father of Jim, Jack, and Johnny Tyone.

Notes on Ahtna Personal Names

Ahtna and other Alaska Athabascans have a unique system of personal naming that has played a central role in their culture. There are sharp contrasts in the cultural uses, functions, and structure of Ahtna personal names as compared with geographic names in Ahtna or in other Athabascan languages.

Among Ahtna speakers personal names are a favorite topic of discussion and reflection. The Ahtna seem to have kept giving children personal names well after historic contact (whereas many Alaska Athabascans discarded this tradition when they were given Christian names). Virtually all of today's Ahtna elders have Ahtna personal names. Ahtna speakers know the personal names of most of their contemporaries. In the past Ahtna personal names were not passed on, but were unique for the individual. The names were given sometimes at birth but more often after a child developed his or her personality. Names tend to be avoided or tabooed after a person is deceased. Consequently, we often find that early historic recordings of personal names from 150–200 years ago cannot be re-elicited.

In 1980 and 1981 Jim McKinley and Martha Jackson summarized some of the rules about the use of personal names. Jim: "You do not address a person by his personal name, but only by his (or her) kin term. Personal names are not supposed to be mentioned in the presence of the named person's clan relatives. After a person dies, a name is not given to another. Chief's names are only to be used in potlatch speeches." Martha: "You never give the name of a deceased person to another person. Parents do not name their own kids. The names are given by a grandfather or uncle or aunt, maybe when the child is about three years old."

In the preceding narrative, Ahtna expert Frank Stickwan lists the personal names of well-known head men from the larger Ahtna villages along the Tazlina River toward Tyone Lake. For two sites, **Bendilden** and **Ben K'atggeh,** Frank listed the names of four head men, which must reach back to the early 19th century. In a short text Katie and Fred John (Kari 1986:21–24) listed sequences of famous chiefs' names for the Upper Ahtna: for four chiefs for Mentasta, two for Suslota and seven for Batzulnetas. Since the Johns were able to estimate the time of death of these chiefs, this set of Upper Ahtna chiefs names allowed us to estimate dates for several early historic events, such as the two times Russians were killed in Upper Ahtna territory in the mid-1790s and in 1848 (Kari 1986:75–87, 107–114).

Until the 1890s most Ahtna did not have "white man" (i.e., conventional or legal) names. The earliest census of Copper River, collected in 1910, contains many persons listed with only an Ahtna name. One common pattern at this time was to name the head of a family for his village—Mentasta John, Mentasta Pete, Gakona Gene—and then to name children with a man's name

(John, Pete, Gene, etc.) as the family's surname. A few Ahtna surnames are based on personal names, such as Eskalida (from **U'eł 'Sc'eldiy' Ta'**) or Goodlataw (from **C'utl'ata'**). Also, some official place names are based upon the names of Ahtna people; for example, Batzulnetas is from an actual Ahtna name, and Ewan Lake is based on the surname Ewan, which is from the Russian name Ivan.

It is rare in Ahtna to find any personal names among the Ahtna place names. Of the eight place names that refer to persons, four use modern names, such as **Tom Neeley Bene'** and **Nikolai Ak'ae.** A few are commemorative, such as **Cuuy Ak'ae,** the site near Gulkana named for the famous midget chief. Two unusual place names that refer to persons are:

> **Hwc'ele' Ta' Ik'e Ngedzeni** 'father of rags is standing upon it' (personal name of Doc Billum) [hill at Department of Natural Resources building in Tazlina]
> **Ba'ane Ts'ilaaggen Tak'adze'** 'spring of someone killed him outside' [1-77, spring on lower Klutina R]

The first name, reported by Frank Stickwan, is remarkable in that it uses a person's formal teknonym—a person is named for a child—with **ta'** 'father of,' the name of the well-known 20th-century Ahtna man, Doc Billum. Since virtually no one else in the 1990s seemed to know this name, one wonders just how and when this place name got coined. The second name, according to Jim McKinley, refers to a person who was killed at Ellamar on the coast. This is a rare Ahtna commemorative place name, but note also that it does not use the person's actual Ahtna personal name, but instead is a circumlocution referring to the person.

An important source on Ahtna personal names by the late Ruth Johns (1986) is a compilation of 127 Ahtna personal names. Ruth Johns' study has a nice balance of men's and women's names, and it spans her generation and one or two preceding generations. Several interesting patterns are shown in the names that Johns collected.

females (64)			
with **naa**	without **naa**		
'mother of'	(various nicknames)		
60	4		
males (63)			
with **ta'**	with VERB-**en**	with VERB+0	NOUN
'father of'	'the one who VERB'	verb with no suffix	
11	39	10	3

Many Ahtna personal names are teknonyms: men with **-ta'** 'father of', and women with **-naa** 'mother of'. While almost all of the Ahtna women in Johns 1986 have the teknonym **naa,** it is interesting that only 11 of the 63 male names have the teknonym **ta'.** This implies that for men **ta'** was more formally bestowed. According to Martha Jackson, a father can be named with **ta'** after his first child is born, but I have not pursued the question of how this was or was not bestowed. Also surprising is that all 39 names with nominalized verbs with the suffix **-en** are for men's names. Not one is for women. Regularly when a verb ends in **-en** the singular animate suffix, it apples to either a man or a woman, as in **zdaanen** 'the one (he, she) who is sitting.'

Most personal names have associated explanations about that person's traits or talents. Ahtna values are reflected in the names: generosity, skill with the hands or in hunting, and diligence. The majority of male and female personal names are verb phrases 'the person who VERB PHRASE'.

Dahwdełc'elen, McKinley George; lit. 'one who tears apart words'
U'eł Kaniiłen, Pete Ewan; lit. 'the one with whom something is happening'
Bets'ulnii Ta', Batzulnetas chief in 1885; lit. 'father of someone is respectful of him'
Takol'iix Ta', 19th-century chief of Mentasta; lit. 'father of daylight over water'

The preceding set of names are full verb phrases. However, often there is ellipsis, where the stem of a verb in the name is deleted. In these cases the phrase has an inferred meaning rather than a literal meaning.

U'eł Nits'ulnaa, Grandma McKinley, Jim McKinley's mother; lit. 'mother of with her someone is VERB'; **nits'ul-** has four
　　verb prefixes without a verb stem. Jim said the inferred meaning is "she loves people."
Beni'ts'ilnaa, Martha Jackson; lit. 'mother of toward her someone's mind is VERB'; **beni'ts'il-** has five prefixes without a
　　verb stem. Martha's explanation of the name is "everyone thinks of her."

Another very interesting feature of Ahtna personal names are "caninonymic names," where men were sometimes named for a favorite dog. The chief in the Lower Tonsina area had the name **Nitggaas Ta'** 'father of he turns grey,' his favorite dog. Fred Ewan states that this was common and was usually a man's second personal name. Fred also remembers quite a few nicknames, which were more informally used than than one's main personal name.

Distinct from the personal names were the unique Ahtna titled chieftainships. Sixteen well-known village sites had titles based upon the place name (Kari and Tuttle 2005:22–25). Jim McKinley makes note of four of these titles in Chapter 1. The eight leaders Frank Stickwan mentions for **Bendilden** and **Ben K'atggeh** would have held the titles **Bendil Denen** 'person of lake current place' and **Sałtigi Ghaxen** 'person of Sałtigi' (a hill just north of Tyone Village); see Figure 2-5.

Thus the rules surrounding Ahtna personal names are completely different from those for geographic names. Personal names are unique for individuals and are not passed on, whereas Ahtna geographic names are shared, to be memorized, and have many rules of meaning, structure, and distribution. We have evidence that the vast majority of the Ahtna place names have been passed on for countless generations.

There is the potential for a really excellent master list of Ahtna personal names. There are many sources, including Ruth Johns' 1986 study and deLaguna's and Kari's field notes. Numerous Ahtna audio recordings mention personal names. Song recordings, especially those by deLaguna, mention many personal names associated with songs. The 1910 and 1920 census lists for Ahtna villages have many personal names, some of which may coincide with names from other sources. Such a project would have enduring value for Ahtna people and their culture.

Photo by J. R. Williams,
Bureau of Indian Affairs.

Figure 2-5. Aerial photo of Tyone Village, Ben Katggeh **on the east side of lower Tyone Lake on May 29, 1954.**

The photographer noted that the only two residents at the time were Jimmy Secondchief and Johnny Tyone. The westernmost Tyone Lake is called **Hwtsuughi Ben Ce'e** 'lowland big lake'.

The hill at the right of the photo is called **Sałtigi.** In August–September 1953 archaeologist William Irving surveyed sites in the Tyone Lake and Tyone River area, locating 11 sites and many artifacts. Irving (1957:43) wrote: "The hilltop affords an excellent outlook over the nearly flat surrounding country, about half of which is thinly forested or devoid of timber. A spot about 30 feet across on the higher of two knobs is largely bare of vegetation: on the exposed but little-eroded till were found 32 implements and an assortment of other cultural debris. Most of the material was found on the southern side of the knob, which suggests use of the site during the winter months when this section would be favored by the low sun." See also Chapter 3, pages 71–73.

Examining this photo, Fred Ewan commented: "**Sałtigi** is a real good luck mountain. There was a village at the base of it, and the old men used to sit up on top of it and watch for game all over. They would holler down to village below when they saw game."

The 60-mile long trail between Tyone Village, Crosswind Lake village (**K'estsiik'e**) and Gulkana was traveled throughout the year. Fred Ewan states that in the winter typically people made this trip in two days, stopping at Crosswind Lake. The summer the route was longer, routing around lakes, and they might take three to four days each way.

3

Saen Tah Xay Tah C'a Łu'sghideł

We Used to Travel Around in Summer and Winter

Jake Tansy

Figure 3-1. Jake Tansy in August 1981 near the Valdez Creek village site C'ilaan Na' **[3-22].** Rusty Hill, **t**he mountain in the background, is **C'enaa Dzele'** 'sign mountain'.

Photo by Priscilla Russell.

Jake Tansy was born in 1906 at Valdez Creek and resided in Cantwell for many years until he passed away in the fall of 2003. Jake was the expert on the uplands of the upper Nenana and Susitna rivers–the traditional territory of the Valdez Creek–Cantwell Ahtna band. I worked with Jake on about 25 occasions between 1974 and 2001. Jake had rare combinations of experience and skill as an outdoorsman and raconteur. Jake's 1982 book (Tansy 1982) is a fine collection of Ahtna **yenida'a** legends. Jake's career Ahtna place name network of about 350 names presents the northwest edge of Ahtna territory and covers the area of the upper Nenana River above the Healy River and along the upper Susitna River to Devil Canyon.

Jake has provided the most detailed travel, land use and place name materials of any Ahtna speaker. Jake's delivery style is rapid and precise. These are masterful examples of topographic description and orienteering with complex interplay of the symbolic and structural features of the toponymy, the riverine directionals system, and great familairty with landscape (geology, hydrology, vegetation etc.) Part 1 was published in Kari 1999:36–39 with an interlinear translation and a trail map.

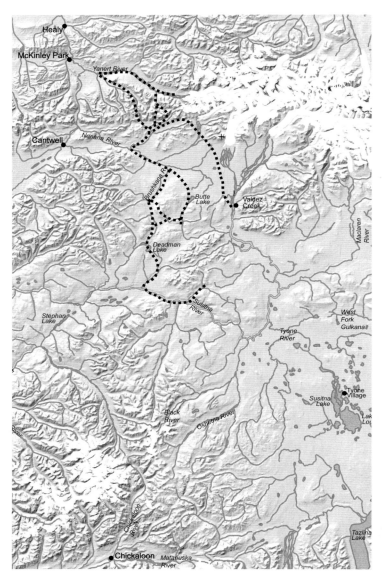

Figure 3-2. Jake's Tansy's intricate routes in Chapter 3. The routes in Part 1 are about 120 miles; those in Part 2 are about 90 miles.

Recorded by JK with Jake Tansy on Nov. 6, 1980, in Fairbanks on AT21(2) or at5006. The first segment was published in Kari 1999. Total: 8:24. Sound file chp3-jaketansy.wav.

(1)
Summer Travel Routes: Mouth of Brushkana River to Yanert Fork to Valdez Creek Village
Starting point: Brushkana River cabin
Ending point: Valdez Creek
Verified routes total about 120 miles.

Xona first nen' ta'stghideł de c'a saen ta c'a Bes Ggeze Na', Saas Nelbaay Na' hwcets'edeł.
/When we first went out in the country during the summer we would ascend 'bare bank stream' [3-1, Wells Ck] or 'sand that is grey stream' [3-2, upper Wells Ck].

Niłdenta hw'eł Dghateni yi 'eł Tanidzeh xu Dghateni ts'idiniłen.
/Sometimes also to 'stumbling trail' [3-3] or 'the one in the middle' [3-4] or the 'stream flowing from stumbling trail' [3-5].

Yic'a Tanidzehi Deyii Na' k'a hwk'e'e kudełdeye dze' xuyii hwtes'sghideł.
/There at 'the canyon of the stream of the one in the middle' [3-6, creek from NW into Wells Ck], a short ways before that, we went through a pass.
0:26
C'eldaan'ne ełe 'unggu xangguxu tes ts'udaełde kiyniziix ts'e'
/Or if some of the people to the upland, next upland wish to go over a pass and

Xangguxu Dghateni hwtes kedeł.
/at 'upland stumbling trail' [3-7, westerly trail to Yanert Fork] they went over a pass.

Ba'aadze' den łu' Tl'ahwdicaax Na' hwts'e' hwcets'edełde Łena'udghidlen xunt'ae.
/over from there to 'valuable headwaters stream' [Yanert Fork, 3-8] we ascend what is 'streams join together' [3-9, Louise Ck].

Yet Tl'ahwdicaax Na' 'usu tayenk'e ghenaay 'eł xona ka'sdal'iix.
/There at 'valuable headwaters stream' [3-8] out on the river plain we would see caribou.
0:45
Niłdenta łdu' yet Tl'ahwdicaax Na' 'udaa'a 'unaa daa'a ts'ets'edeł dze'
/Sometimes then there we come out downstream and across and downstream of 'valuable headwaters stream' [3-8, Yanert Fork] and

dae' Nts'ezi Na' hwts'e' tes ninats'edeł.
/that way we come back through a pass to 'nts'ezi stream' [3-10, Moose Ck].

Nts'ezi Na' ye cu tcenyii kughił'aen', I mean dahtsaa, dahtsaa hwghił'a'.
/At 'nts'ezi stream' was an underground cache, I mean there was an elevated cache.
1:05
Teye k'a 'udii c'etsen' nghiłggaasi dahtsaa t'anahghilaeł.
/All the time they put lots of dry meat in the pole cache.

Xona ye łu Nts'ezi Na' ye kae na'sdelgges dze'
/Then there we would come back with that (meat) on 'nts'ezi stream' and

dets'en dets'en Nts'ezi Na' ba'aa dghilaay ghakudaan de kanats'edeł.
/that side, that side beyond 'nts'ezi stream' we would ascend back up through a canyon in the mountains.

N'eł Bes Ggeze Na' ye cuu cu Bes Ggeze Na' hwk'e koodaan yehwk'e na'sdelgges xu.
/Or at 'bare bank stream' [3-1, Wells Ck] there also is a gorge and we would come back through where the canyon goes through 'bare bank stream'.
1:28
Bes Ggeze Nangge' na'stedeł dze'.
/We would get back to the 'uplands of bare bank stream' [3-11, Wells Ck uplands].

'Unggu Saas Nelbaay cu ye łu udi'aan cu ye same c'ena' su, cu 'ungga cuts'en dze' nay'det'aan.
/In the upland of the one named 'sand that is grey' [3-2, upper Wells Ck], it is the same stream but in the uplands (the fork) has a different name.

Saas Nelbaay Na' Ngge' cu ye xona ba'aa Łuyinanest'aani Na' su hwtah dadaa' kanats'edeł.
/'Uplands of sand that is grey stream' [3-12, upper Wells Ck uplands] then there again beyond there we would ascend 'stream of the one protruding into the glacier' [3-13 upper Nenana R] or sometimes toward the downstream.

Łuyinanest'aani Na' yanaasts'en hwtl'adaak'e su Ts'es Ce'e de gaa hwnax gaani 'idighiłcaax xu dez'aan.
/On the other side of 'stream of the one protruding into the glacier' is a bluff 'big rock' [3-14, rock bluff above Siksik L] that is as large as this house.

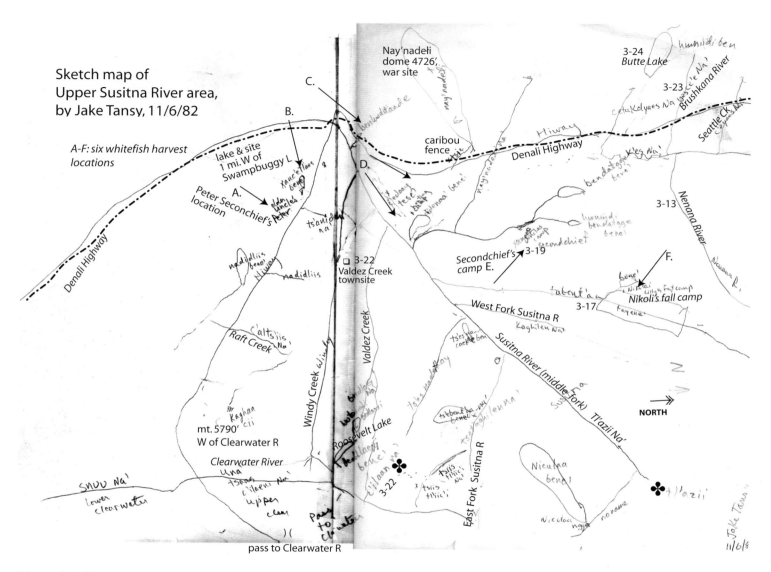

Sketch map of
Upper Susitna River area,
by Jake Tansy, 11/6/82

*A-F: six whitefish harvest
locations*

Figure 3-3. Sketch map of Upper Susitna River area, by Jake Tansy, in November 1982. This area at heads of the Nenana and Susitna rivers was Jake's home territory. The original map is very cryptic with my shorthand notes. When labeled as in Figure 3-3, we can see the brilliant details that Jake has imparted. There are 39 place names, about 10 sites for whitefish or for hunting, and several trails and passes. Seven names on this map are mentioned by Jake in Chapter 3 and two were mentioned by Dick Secondchief (see Figure 3-7).

Ye su xona 'udii hw'eł hnats'at'iix hwghak'aay hw'eł łu'steltset c'a snakaey ts'ghile' de.
/We always used to play there, we would run around on the flank (of the rock) when we were kids.
2:00
Yak'a k'adii c'a dae' z'aan.
/It is still there like that.

Xona yet łu' ye c'a ye łu Łuyinanest'aani Na' tsen Saen Tene na'sghideł.
/Then there at 'stream of the one protruding into the glacier', we come back to the lowlands to 'summer trail' [3-15, main trail at base of range].

Ye yak'a Łuyinanest'aani Na' tene ka'sghideł hwna.
/There at 'stream of the one protruding into the glacier' we ascend the trail.

Xu yae' Kuyxi Dghilaay Cene 'ane Kuyxi Dghilaay Cene ba'aa Tabenł'aa Tayene' yi na'sdaldeł.
/And this way we descend again over to 'base of marmot mountain' [3-16, base of mountain between West Fork Glacier and Nenana R] and over to 'base of marmot mountain' to 'lake current flows-river plain' (3-17, West Fork glacier plain).

Tabenł'aa Tayene' du' four miles ghiłnaes xunt'ae, four mile wide ce'eł nlaen.
/'Lake current flows-river plain' is four miles wide, it is four miles wide.

K'ay' k'ali' ukedi'ah, t'ae' hwtsicdze' de gaa airplane field k'e sunt'ae de.
/There are no willows protruding on it and it is all just like an airplane field.

Yełu' 'utsii ye c'a 'utsii ye Tabenł'aa Tayene' 'utsii taz'aa de Tabenł'aa Bene' hwdi'aan.
/And downland from there downland of 'lake current flows-river plain' in the downland is situated the water called 'lake current flows-lake' [3-18, lake off West Fork of Susitna R].
2:28
Tabenł'aa Bene' gha yet ts'inats'edeł.
/We come back out at 'lake current flows-lake'.

Xona yet xu tsene łyes kanats'edeł dze',
/Then in the lowlands we ascend again in the dwarf birch (buckbrush), and

ye Ben Datgge cu yae' ts'inats'edeł dze' Ben Datgge Na'.
/there at 'lake up above' [3-19, lake below West Fork of Susitna] we come back out that way to 'stream of lake up above' [3-20, creek into Susitna R].

Hwtsene xona yełu' 'utsiit Ts'itu' gha.
/In the lowlands then there by 'major river' [3-21, Susitna R].

Xona some, bede c'a kekon' . . . kekon'sghiyeł, cu kałetdiłdox hwna cenuu negha kekaes.
/Then someone, *f.s.* someone would build a fire and while the smoke ascended, they would come over to us in a canoe (from Valdez Creek village, C'ilaan Na', 'abundance stream,' 3-22).

Dae' su tkat'aen', xona.
/That is how it was.
3:12

(2)
Summer or Winter Travel Routes: Brushkana Cabin to the Middle Susitna River
Starting point: Brushkana River cabin
Ending point: Watana River on the middle Susitina River
These routes total about 90 miles.

1922 tah yehwts'en su my brothers my mother 'eł łu Brushkana xuhwts'e' 'sghideł.
/After 1922 we would go with my brothers and with my mother to Brushkana [3-23].

Yehwtseh hwgha Tl'ahwdicaax Na' yae' łu'sghideł.
/Previously [I spoke] about how we would go that way to 'valuable headwaters stream' [3-8, Yanert Fork].

K'adii łu Brushkana xuhwts'e' 'snidaetl' tah xugha Hwniidi Ben xu yae' 'stedeł dze'
/Now we went to Brushkana R and then we would go over to 'upstream lake' [3-24, Butte L].

Bes Ce'e Na' hwcets'edeł.
/and would go upstream along 'big bank stream' [3-23, Brushkana R].

Brushkana k'adii k'a 1926 yet hwnax 'skułtsiin.
/Now at Brushkana we built a house in 1926.

Yehwts'en c'a 'udii every summer k'a 'unaa hwts'enats'edeł dze'
/From then on every summer we would go out across from there and

yełu yae' da'endze' ba'aa 'udaa'a Kacaagh xu xuhwts'e' łu'stedeł.
/then this way from over there to other side and downstream to the 'large area region' [3-25, Deadman L area].

Yii gha dghilaay nen tah tseles yii gha 'aeł 'eł ts'ghila', ggaał 'eł.
/We had traps set and snares on the mountain sides there for ground squirrel.
3:59
Ye łu ndaan c'a ghanaay c'a negha yaas cuu yi c'a 'sdzghiłghaes.
/There wherever a caribou would come to us also we would kill it.

Figure 3-4. Hwnidi Ben **Butte Lake [3-24], looking south from the northeast end of the lake.**

Both Butte Lake and Deadman Lake [3-25] are clearly labeled on the 1839 Wrangell map as "Knitu ben" and "Kachobena" (Kari and Fall 2003:85–87). Archaeological sites on Butte Lake range in dates from 5,000 to 120 years B.P.

Photo by Priscilla Russell.

Yet c'etsen' ggan ts'ełtsiis c'a xona.
/We would make dry meat there then.

Bes Ce'e Na' yet tseh ya'a hwnax 'skuł'aan dze' ninats'elyaes.
/There to 'big bank stream' [3-23, Brushkana R] where we had previously built a house, we would put everything.

Yeł xona yehwna nahw .. nahwdek'as 'unaane una' nunyeggaay day'tneghał hwna yic'a 'aeł nits'elaes.
/Then while it *f.s.* became cold weather across the stream while the season was open, we set traps for fox.

Yet da'andze' Hwniidi Ben ba'aax Hwniidi Ben gha yic'a 'aeł gha hwnax 'skughił'a'
/And over from there at 'upstream lake' [3-24, Butte L] out from 'upstream lake' we had a house for trapping.
4:28
Ba'aa yehw Hwniidi Ben gha ye na'stedeł xu 'aeł nits'elaes.
/We would go out to 'upstream lake' and set traps.

Yet 'udaa' Hwneldeli Cene da'aa 'stedeł dze' 'udaa'.
/Downstream from there at 'base of the one that is red' [3-26, base of mountain on Wickersham Creek] we would go out and downstream.

Slat Creek yeł dae' hwdi'aan koht'aene k'ali'i c'a 'sdi'ah.
/That place called "Slat Creek" [3-27] we do not have a Native name.[1]

_____

[1] Jake later supplied the name as **Kacaagh Ediniłeni.**

Figure 3-5. Glacier Lake. Glacier Lake in the Delta River drainage is called **Niłyiits'i Bene'** 'skinny lake'. This is in the Tangle Lakes Archaeological District. There are about 46 Ahtna place names in the Delta River drainage. The northern Ahtna boundary was in the area of Trims Camp.

Photo by Suzanne McCarthy.

4:41
Slat Creek dae' hwdi'aan dze' yet cu tent frame 'sghił'a'.
/We had a tent frame at the place called Slat Creek.

Cu ye yaen' na'snedeł dze'
/Also there we always would spend the night there and

satggan łu' da'aa Kacaagh datsiits'en Kacaagh datsen 'utsene hwdaaghe ka'sghideł.
/in the morning beyond there on the lowland side of 'large area' in the lowland, to the distant outlands of below 'large area' (3-25, Deadman L area), out lowland we would climb up above timberline.

Cetakolyaes Na' ye tl'adaak'e cu ye cu tent frame ts'eghił'a'.
/At 'things are carried down to the base-stream' [3-28, lower Deadman Ck] on a bluff we had another tent frame.

Yet yet cu na'snedeł dze', satggan łu 'unaaxe Txastsaet Na' xu 'aeł 'eł hwtsecdze' 'sghila' dadaadze Kacaagh xu n'eł.
/We would spend another night there and in the morning in the area across at 'water level drops-stream' [3-29, creek into Watana R] we had all traps to the area from downstream of 'large area region' [Deadman L area].

Hwtsecdze' hwtsecdze' 'aeł xu nits'elaes xona k'edze' dae' Bes Ce'e Na'.
/We would bring all the traps then back that way to 'big bank stream' [3-23, Brushkana R].
5:20
Ts'i'aadze' yae' gaa Slat Creek ts'en cu Bes Ce'e Na' hwts'e' xu hwghaaghe kakudaan.
/Directly over this way here at Slat Creek side again to 'big bank stream' near there is a gorge.

Udenaa' Na' hwdi'aan.
/That is called 'mineral lick stream' [3-30, creek into upper fork of Brushkana R].

Yi na'aadze' xona nats'edeł cuu Bes Ce'e Na' nats'edeł.
/From close over there we would come back to 'big bank stream' [3-23, Brushkana R].

Bes Ce'e Na' łu gaa snaan 'ene 'eł hwtsicdze' tsets ku'eł nina'sdelaes.
/At 'big bank stream' here we put up all the firewood with my mother and the family.

Gaa yak'a ba'ane Hwniidi Ben xu cu 'aeł ta nase su cu.
/From here over to the other side to 'upstream lake' [3-24, Butte L] we would go for trapping out there as well.

'Udaa' xu k'a tsecdze' Cetakolyaes Na',
/Downstream the entire way to 'things are brought to base stream' [3-28, lower Deadman Ck],

hw'eł yae' 'unaax Txastsaet Na' hw'eł tsicdze' 'aeł ta na'snghideł dadaadze' Kacaagh xu c'a.
/or all the way across to 'water level drops stream' [3-29, creek into Watana R] we would camp out among the traps as well as downstream from 'large area' [3-25, Deadman L area] as well.
5:50
Xona yak'a tsecdze' 'aeł na'snel'iix dze' cu yak'a cu Bes Ce'e Na' nats'edeł
/So thus we would keep checking all the traps and then again we would return to 'large bank stream' again.

Cu dog team tiye dahwdighitey' dze' łi'kaey kae su, łi'kaey t'a'sdel'iix.
/It is quite difficult to explain how we used the dogs, the dog teams.
6:05
Ghanaay łu yeden łu' teye ghanaay c'ghile' 1928 'eł '29.
/At that time there were lots of caribou in 1928 and 1929.

Teye c'a ghanaay 'snghił'aen'.
/We saw a great many caribou.
6:16
Nuu Zdlaade Txastsaet Caek'e xun'e about 10 miles c'a su kulaen ye kole ye 10 miles might be.
/'Where there are islands' [3-31, islands on Susitna R above Watana mouth] is about 10 miles from 'water level drops mouth' [3-32, mouth of east sidestream of Watana R] or less, it might be less than 10 miles.

Yet su tsiis danadze' desuuni c'ighile'.
/There was some really nice ochre there.

'Udii some sometimes cu hwghaaghe 'stayaas hwna yak'a tsiis gha 'utsene hwda'sdiyaa.
/Whenever we would go near there we would go on down lowland-ward for ochre.

Tsiis ts'ughines.
/We get some ochre.
6:48
Yiigha su yełu c'enuu' 'use tazdlaa xu tkut'ae yii xangga k'a tsiis c'ilaen.
/Out from there are the islands are out in the water and upland of there is the ochre.

Yełu' Nuu Zdlaade cu hwdi'aan.
/So that place is called 'where there are islands' [3-31].

Ye xona 'eł kanii ye kudełdiyede su Sdaa Yizdlaayi hwdi'aan.
/From there the next place upstream a short ways is named 'points that are in there' [3-33, point and hill on N bank of Susitna R].

Hwyii 'un'e łu Cets'i Caek'e, Cets'i Caek'e cu
/Within area (a canyon) to the upstream is 'spearing mouth' (3-34, mouth of Kosina Creek on the S side of Susitna R), also

hwc'aats'e hwghaaghe su Nac'alcuut Na' hwdi'aan.
/opposite there and nearby is called '(sth.) food is brought back stream' [3-35, Jay Ck].
7:17
Yet cu yet Nac'elcuut Na' 'uniits'e dae' dae' Txastsaet Na' xuhwts'e' Ts'anit'ehi Na' hwdi'aan.
/There also there down from the upstream of '(sth.) food is brought back stream' [3-35, Jay Ck] that is 'water level drops stream' then to a place called 'that which is rough-stream' [3-36, creek from N into Susitna R].

Dghilaay Ts'anit'ehi gha 'ey'dełeni gha.
/That (stream name) is by 'mountain that is rough', it is flowing from [3-37, mountain 5046' N side of Susitna R].

Cu yedets'en łu deghilaay ghene kulaen tes ce'e i'deltsiin.
/Also from there in a curved slope on that mountain are shaped big hills.
7:33
Ts'abaeli tahwts'en cu 'utgga dghilaay ts'e' ta tes ce'e tnezdlaa xu tkut'ae.
/The spruce timber also is up to the mountain there where the 'big hills' are.

Nac'elcuut Nelyaade udi'aan.
/Those are called 'the ones that lay at *(sth.)* food is brought back' [3-38, ledges, buttes on slopes of mountain 5046'].

Yeden cu ye cu cu yedets'en Txastsaet Na' hw'eł Nac'elcuut Nelyaade cu hwghatgge łu tcenyii kughile'i 'eł.
/There on the side of 'water level drops stream' [3-29, stream into Watana R] and in between 'the ones that lay at *(sth.)* food is brought back' [3-38, hills on that mountain] there was an underground log cache.
7:52
Ye su tcenyii tcenyii kughile'.
/There was an underground cache.

C'etsen' nen' kughelyaa.
/There where meat was stored in the ground.

Yet su 'utgga dghilaay yet Txastsaet Na' 'eł hwnez'aani Nacez'aani udi'aan.
/And up the mountain there at 'water level drops stream' [3-29] is standing the one called 'our heart' [3-38, mountain 4120' E of Watana R].
8:06
Nya' k'a 'eł dae' Txastsaet Na' 'eł łkents'edeł dze'
/Going on that way we would cross 'water level drops stream' [3-29, stream into Watana R], and

Debetse' Na' łkenats'edeł dze' Cetakolyaes Na' xona.
/we would cross 'sheep head stream' [3-40, Watana R] and then to 'things are brought down to base stream' [3-27, lower Deadman Ck].

 Xona huh
/That's all.
8:20

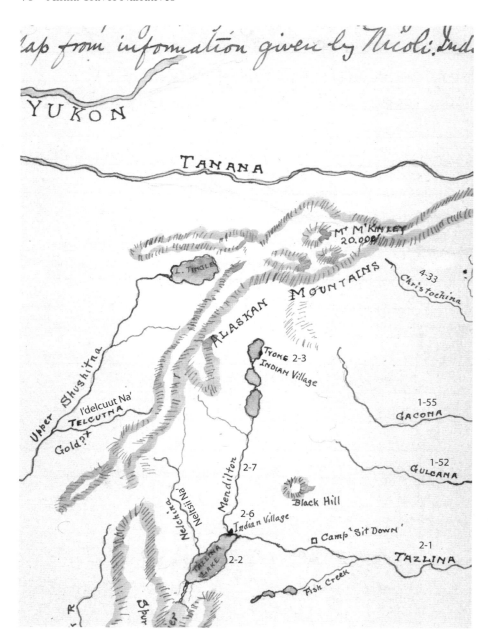

Figure 3-6. Small-scale sketch map by Basil Austin done in collaboration with Chief Nickoli of Tazlina Lake.

Compare this with Austin's map (Figure 2-3) centered on the Tazlina and Klutina rivers, an area that Austin and his partners had familiarity with in 1898–99. Reference numbers for places in Chapters 1, 2, and 4 are noted.

Information for this map was imparted by Nicoli of Mendeltna village. Interesting features are Mendeltna Creek as the route to Tyone Village; an indication of the Tanana River; the heads of several rivers that lead up to the Eastern Alaska Range; and the name "telcutna", which is the Ahtna pronunciation of the name for Talkeetna River. Therefore Nicoli had firsthand knowledge of the most of the upper Susitna River drainage as well as the central Copper River country.

Cultural Geography of the Hwtsaay Hwt'aene,
'the small timber people'

The term **Hwtsaay Hwt'aene** applies to all of the Western Ahtna, those at Tyone Lake as well as the Valdez Creek–Cantwell band. In 1953 archaeologist William Irving worked with Jimmy Secondchief, who provided him with important information about the **Hwtsaay Hwt'aene.** Irving wrote (1957:39–40):

Generally speaking, on a year round basis, the country is not rich enough in these food resources to sustain large villages. On the basis of the recent houses discovered and information from local informants, it would appear that the early post-contact population was no more than one hundred persons. Presumably, these were living in scattered groups of five to thirty individuals following a semi-nomadic existence.

From Jimmy Second Chief, a willing and able informant, the following was learned. The annual cycle was divided into two major phases, dependent upon the feasibility of fishing. From midsummer through December, the principal activity was fishing. The group at this time would accordingly locate near spots suitable for using "V" and basket [fish] traps. Caribou and moose would be killed from time to time throughout the year, but were given particular attention in late summer and early fall. At this time bulls were fat and skins most suitable for clothing. Fish however formed the most important food item.

By midwinter, shallow places in the lake would freeze to the bottom and fishing would no longer be profitable. By this time also the meat stores from the previous fall would be exhausted. It was then necessary that extensive hunting of moose, bear, and beaver be carried out. . . . Moose and caribou fences, in conjunction with snares and the surround, were used. This would continue until breakup, after which the hunters would go into the hill country often as far as the Talkeetna Mountains. Here they would remain hunting caribou until midsummer, when they returned for fishing. Travel was usually on foot; infrequently by canoe. . . . Vestiges of an elaborate system of trails may still be seen, and even now foot travel for distances of forty or fifty miles is routine. The Tyone River people, then, are part of a cultural subgroup indigenous to the intermontane region and think of themselves as different from the river people of the Copper and lower Susitna rivers.

Other sources on the Western Ahtna include a walking tour of the Tyone Lake–Knik trail by Jim Tyone in 1912 in which he highlights 33 place names along the 170-mile trail, and further details on the Western Ahtna/Dena'ina interface in Kari and Fall 2003:223–225 and Chapter 10.

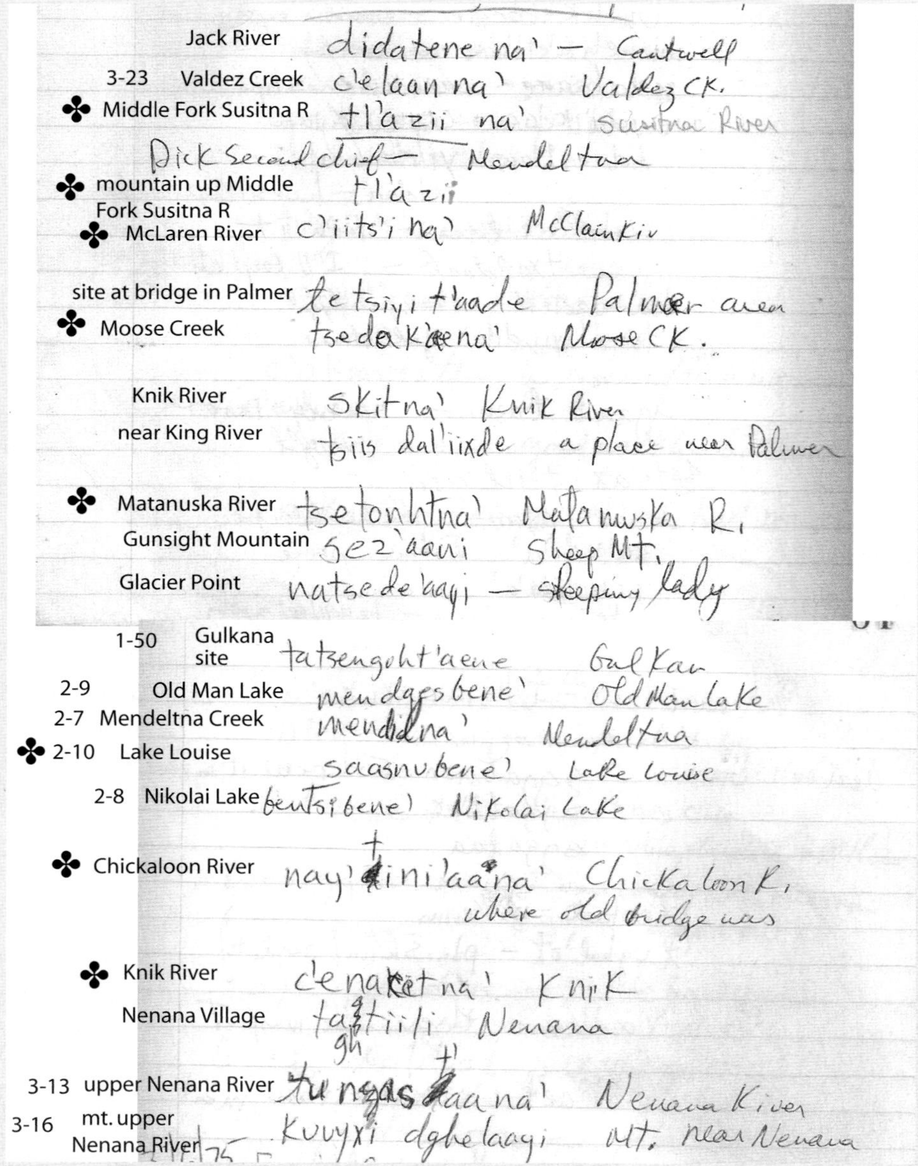

	Jack River	didatene na' —	Cantwell
3-23	Valdez Creek	c'elaan na'	Valdez Ck.
❧	Middle Fork Susitna R	tl'azii na'	Susitna River
		Dick Secondchief —	Mendeltna
❧	mountain up Middle Fork Susitna R	tl'azii	
❧	McLaren River	c'iits'i na'	McClaren Riv
	site at bridge in Palmer	te tsiyi t'aade	Palmer area
❧	Moose Creek	tsedak'aena'	Moose Ck.
	Knik River	skitna'	Knik River
	near King River	biis dal'iixde	a place near Palmer
❧	Matanuska River	tsetonhtna'	Matanuska R.
	Gunsight Mountain	gez'aani	Sheep Mt.
	Glacier Point	natsede'aayi —	sleeping lady

1-50	Gulkana site	tatsengoht'aene	Gulkana
2-9	Old Man Lake	mendaesbene'	Old Man Lake
2-7	Mendeltna Creek	mendeltna'	Mendeltna
❧ 2-10	Lake Louise	saasnubene'	Lake Louise
2-8	Nikolai Lake	bentsibene'	Nikolai Lake
❧	Chickaloon River	nay'dini'aana'	Chickaloon R, where old bridge was
❧	Knik River	c'enaketna'	Knik
	Nenana Village	taltiili	Nenana
3-13	upper Nenana River	tungas'aana'	Nenana River
3-16	mt. upper Nenana River	kuuyxi dghelaayi	mt. near Nenana

Figure 3-7. Field notes with Dick Secondchief, March 5, 1976 (Kari notebook #3:60–61).

Secondchief listed 22 names, eight of which are mentioned in Chapters 1, 2, and 3. The symbol ❧ marks features that are common to Secondchief's list and the 1904 Moffit map. See also Figure 3-3, Jake Tansy's sketch map.

On March 5, 1976, Dick Secondchief of Tyone Lake and Mendeltna stopped by to visit Martha Jackson at her home in Copper Center while I was working with her. I had not spoken with Dick before. He was in his early 80s, and I knew he was an expert on **Hwtsaay Hwt'aene** geography. Prompted by a few questions, in about 20 minutes Dick gave this list of 22 place names. Consider the geographic spread of places Dick had been to, mainly by foot travel: in the east a site near Gulkana; in the north, Nenana Village; in the south, a name for Knik River. (If plotted as a triangle, this is about 300 miles on the west, about 180 miles north-to-east, and 200 miles south-to-east.) All 22 of these names have independent confirmation by Jim Tyone, Jake Tansy, John Shaginoff, and others.

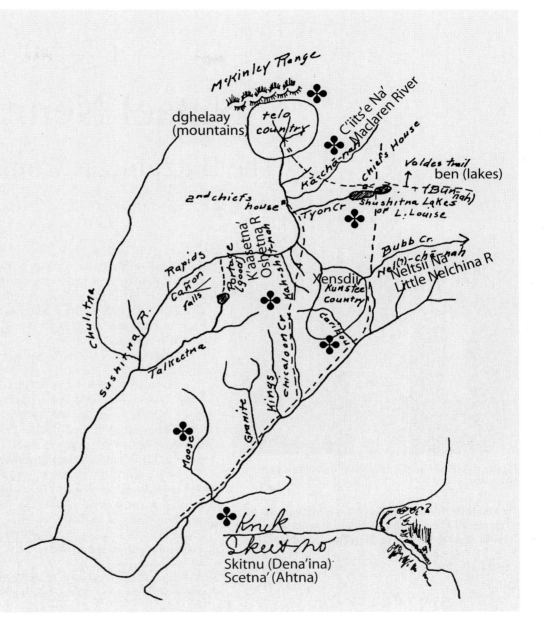

Figure 3-8. 1904 sketch map by Fred W. Moffit.

The 1976 notes from the brief encounter with
Dick Secondchief have real significance when
compared with a 1904 sketch map obtained by
geologist Fred W. Moffit.

This map must have been provided by a person
with a sense of the regional geography very
similar to that of Dick Secondchief. The map
traces the entire drainages of the Matanuska
and Susitna rivers, with more than 300 miles of
intersecting trails and is marked with six Ahtna or
Dena'ina place names written by Moffit. The symbol ✤ marks features that are common to Secondchief's list. Also note "2nd chief's house" marked
near mouth of McLaren River (**C'iits'e Na'**), which
was also noted by Secondchief in 1976. We
can conclude that both of these Western Ahtna
experts knew vast details about the uppermost
Susitna River, the Tyone Lake country, and the
Matanuska River trail system. In Kari and Fall
(2003:231–232), other details on the 1904 Moffit
map are explicated.

4

Natael Nenn'

The Batzulnetas Country

Katie John

Figure 4-1. Fred and Katie John in Mentasta in 1983.

The mountain in the background is **Mendaes Dzele'** 'shallows lake mountain'. As chief of Mentasta, Fred John Sr. held the chief title **Men Daes Ghaxen.**

Photo by James Kari.

Katie John of Mentasta is a **kuy'aat**—a female chief—and leader of the **Tatl'ahwt'aene**, 'the headwaters people,' the Upper Copper River Ahtna. Katie John was born at Slana in 1915, the daughter of Sanford and Sarah Charley. Sanford Charley was the titled chief **Stl'aa Caegge Ghaxen** for the Slana area from about 1910 to 1940. Katie married the late Fred John in 1938, and they moved to Mentasta, where they raised 10 children.

Natae̱lde, or Batzulnetas village on Tanada Creek, has been the locus of an array of important events—legendary, prehistoric, historic, and contemporary (Simeone 2009).[1]

Alaskans came to know Katie John during in the landmark subsistence rights case, *Katie John vs. State of Alaska*. During the course of the case (1984 to 1997), Katie John spoke from her memory and from her heart about Ahtna law and tradition. One memorable statement by Katie John is, "Everything I know I keep in my head." Supporters of Alaska Native subsistence rights use the following phrase on T-shirts and bumper stickers: "Don't mess with Katie John."

I have worked with Katie on numerous occasions dating from 1974. Her assembled recordings are about 10 hours in length and cover a wide range of topics. This narrative, published in Kari 1986:181–193, was recorded June 13, 1981, as we drove in my car from Mentasta to Slana, stopping to view several places. Sounds of passing vehicles can be heard occasionally. This is the best Ahtna woman's travel narrative on record. Katie John's third segment overlaps quite a bit with Adam Sanford's fifth segment.

---

[1] Locally called "Banzaneta," the name Batzulnetas was recorded by Allen in 1885 (Allen 1887:67) based upon the personal name of the chief **Bets'ulnii Ta',** 'father of someone is respectful of him.' See Figure 4-3.

Recorded on June 13, 1981 on AT29 or AT0810. Sound file chp4-katiejohn.wav.
Total 12:41

(1)

Starting point: Batzulnetas
Ending point: Mentasta
Route length totals about 25 miles.

Netsehtah koht'aenn 'iinn gaa gaa nen' k'et łuxac'ehdelyaayi gha łuhnidaetl.
/Before us, the people on the land here used to go hunting in groups.

Mendaes hwt'aenn 'iinn,
/The 'shallows lake' [1-72, Mentasta] people,

Bes Ce'e hwt'aenn 'iinn 'eł niłt'aay datsiit Kolgiisde.
/and the 'big bank' [4-1, New Suslota village on Suslota Creek] people meeting downland at Kolgiis [4-2, Bear Valley Ck].

Kolgiis Na' datsiists'en de yihwk'e niłt'aay kadeł dze'
/They used to meet on the lowland side of 'Kolgiis Creek' and
:32
Yehwts'en danggexdze' Mendaes hwt'aenn 'iinn łuxac'ehdelyiis dzen 'uka.
/From there from the uplands the 'shallows lake' people used to hunt for muskrat.

Bes Ce'e hwt'aenn 'iinn datsen.
/The 'big bank' people (used to hunt) in the lowlands (on the lower Slana R).

Niłghakedeł de niłdze' nixu' niłehdetniix.
/When they met together, they would talk to each other.

"Dae' ts'en nts'eniix naxacdghostiis," niłehdetniix.
/"We say to you that I'm going to hunt on this side," they would tell each other.

"Agha'," niłehdetniix dze' xu' k'et'iix.
/"Okay," they would tell each other and it would happen like that.
0:57
K'alii den dae' Sasluuggu' hwt'aenn 'iinn k'alii Mendaes hwt'aenn 'iinn xunen' ka'iyaal dze'.
/Neither the 'small salmon' (1-70, Old Suslota) people nor the 'shallows lake' people would go up into their (i.e., the others') country.

Yae' dzen li'i kesghiige.
/They could not kill muskrat that way (without permission).

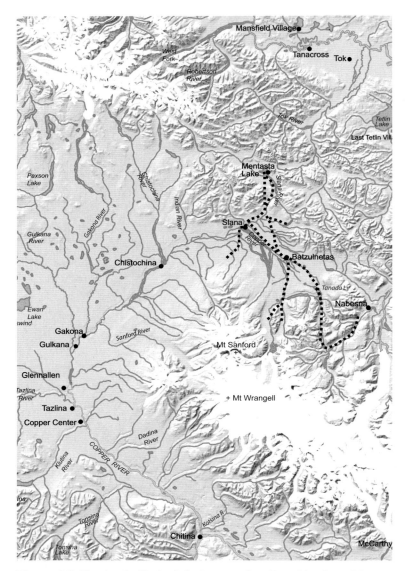

Figure 4-2. Routes in Katie John's narrative. Part 1 is about 25 miles; Part 2 is about 80 miles round-trip; and Part 3 is about 35 miles.

Niłghakedeł dze' xu' niłehdetniixi gha' yaen' koht'aenn 'iinn xu' dghat'aen'.
/They met together and the people did just what they told each other.'
1:17
Datsuux yae' igha yanidan'a hwts'en koht'aenn 'iinn tene kulaen de.
/Down in the lowlands for it (game) from the ancient times there are Ahtna trails.

'Utsii Nataełde hwts'en k'a 'utsiidze tene kulaen dze'
/From downland at 'roasted salmon place' [1-69, Batzulnetas village] there are trails coming up from the lowlands.

Sasluuggu, Bes Ce'e kedi'aan.
/'Small salmon' [1-70, Old Suslota], and 'big bank' [4-1, New Suslota on Suslota Ck] they are called.

K'adii yedu' Suslota kehdił'aan.
/Now they have called that 'Suslota.'
1:33
Bes Ce'e c'a kedi'aan ya xuhwk'e kudeł dze'.
/They would pass by also what is called 'Big Bank' and.

'Utsiidze htediił dze' all tene xu niłtankuz'aa.
/Coming up from below, all the trails joined together.

Duu kohtaenn 'iinn nenatseh ts'en koht'aenn 'iinn tene.
/Around here are the people's trails, from the people before us (our ancestors).

Ngga Mendaesde yet gaa kekudeł dze' hdeke' kaen' tene kulaende.
/They would pass by here upland at 'shallows lake place' [1-72, Mentasta], going on their own feet, where the trails are.

Sii c'a sc'aen ghałe' xuh łunesyaa dze' yii gha' 'eł estnes xuh tene.
/I too went around there when I was a child, and that is how I know the trails.
2:00
Five years old 'ełaen 'eł xuh łunesyaa sta' snaan 'iinn 'eł.
/When I was five years old I went there with my mother and father.

Figure 4-3. This 1905 photo of Mentasta village Men Daesde **[1-72] shows details of buildings and construction.**

In the foreground is a weir across **Mendaes Na',** the outlet of Mentasta Lake. The mountain on the right, on the north side of Indian Pass, is called **Tuu Ts'eni,** 'the one on water side.'

Photo courtesy of Geoff Bleakley.

Xu łusnidaetl dze' all xutsiidze tene koht'aenn 'iinn tene sii c'a 'eł 'estnes.
/We traveled around and I too know all the people's trails from the lowland direction.
(break)
2:26
Datsuux yae' tl'ogh Tl'ogh Tngełnaay da . . . danaa Bes Ce'e dze' tene kehwdi'aa de.
/Down in lowlands at *f.s.* 'grass that moves' [4-3, place on trail to Suslota near mi. 70 on Tok Cutoff] across the way from 'big bank' [4-1, New Suslota] the trail extends.

Yet ca tl'ogh . . . Tl'ogh Tngełnaay kedi'aan.
/That place is called *f.s.* 'grass that moves'.

Ye xungge c'ena' Slana, Stl'aa Caegge Na' Ngge' htediił dze' xatl kaen' xey tene.
/There upland from the Slana River, they would go to 'rear mouth river uplands' [4-4, lower Slana R uplands] with sleds on a winter trail.

'Unggat Xoos Ghadl Zdlaa keniide,
/Up at where they call it 'horse wagons are there' [4-5 on Eagle Trail N of Bes Ce'e],

ye danggasts'en ta xona danggehdze dae' tic'akedeł.
/there then they would go out into the country to the upland side or from the uplands.

Ba'aa yet Tak'ez'aann gha det'ax 'en tah 'en tah tene kughile' dze'.
/On other side is 'object in the valley' [1-71, hill at Indian Pass], by the pass a trail went through the pass.
3:06
Xu'en ta kakedeł dze' bayggat Tes T'aa Menn 'ehdideł dze'.
/They would go up to the other side and they would come out below 'lake beneath the hill' [4-6, 'fifteenmile lake', north of Indian Pass].

Tes T'aa Menn' Tes kaxatl . . . xatl 'eł kekudeł dze'.
/They would *f.s.* pass on sleds by 'hill of lake beneath the hill' [4-7, hill near 'fifteenmile lake'].

'Unaa Tes T'aa Menn' Ts'ediniłeni yii yii na' ngge' take . . . kakedeł.
/Across where a stream 'one that flows from lake beneath the hill' [4-8, stream from hill near 'fifteenmile lake'] they would go upland there on that creek.

Yihwts'en xunaann' htedeł dze' Taggos Menn' 'ehdideł dze'.
/ From there they would go across and they would come to 'swan lake' [4-9, lake north of 'fifteenmile lake'].

Taggos Menn' Tes cu ka . . . kekudeł dze'
/They would pass by 'swan lake hill' [4-10, hill between lake and Slana R] and

Yihwts'en xona K'ekotceni yii.
/then from there into 'on the flats' [4-11, flat between lakes and Slana R].

K'ekotceni k'et naann' tene kughile', xuhgha xatl tene, xey tene.
/At 'on the klats' there is a trail across the way by there, a sled trail, a written trail.

Yet bayggat xona Stl'aa Na' na' keniide
/Below there, then, at what is called 'rear river' [4-12, Slana R], at where they call

Nacox T'ax keniide gha ye tah ke .. kekedeł dze' hungge Nacoxt'ax Na' Ngge' htedeł dze'
/by 'slough pocket' they say, they come there and they start upland to 'slough pocket creek uplands' [4-13, slough or west side of Slana R near mile 70 of Tok Cutoff] [4-14, uplands of stream into Slana R]

'Unggat xona Stl'aa Na' kadighiłen de yet kekedeł dze'.
/They go upland there where 'Rear River' [Slana R] flows up and out.
4:15
Yihwts'en ngge' xona htediił dze' 'ungga Mendaesde.
/From there they would go upland and upland to 'shallows lake place' [1-72, Mentasta].

Xutsiidze de, Mendaes Caegge yetah xona ketakedeł dze'.
/From the lowlands they would pass 'shallows lake river mouth' [4-15, where Mentasta L outlet meets Slana R].

Mendaesde xona Tak'ae yet kedeł.
/'Shallows lake place' then they would come there to 'timbered valley.' [1-72, Old Mentasta; 4-16, valley area in the Mentasta Lake/Mentasta Pass area].

(Continued on page 82)

Solved: The Origin of "Tok"

For some time Shelly Marshall of Tok has been researching sources, folklore, and theories as to the origins of the Alaska town named "Tok." In two articles in the Tok *Mukluk News* (December 3 and 17, 2009), she carefully reviewed early maps, noting various early spellings for Tok River such as Tokai River (Allen 1887), Tokna River (Bourke 1898), Tokioi River (Powell 1901), and the one-syllable name Tok River (Brooks 1900, Schrader 1902). She also noted that Allen's placement of the earliest name, Tokai River, appears to be more like the course of the Little Tok River, starting somewhere just east of Mentasta Pass.

Marshall noted this passage in Allen's report from June 6, 1885. After extensive discussions with people at Mentasta about trails and geographic features at the divide between the Slana and Tanana rivers, Allen (1887:72–73), apparently in the Mineral Lakes area, wrote:

The headwaters of the Tokai River are not more than a mile or two from the lake (Mentasta). This so-called water shed is in reality a pass, 800 to 1,500 feet lower than the mountains on each side that are barren of everything save a little grass, spruce and much moss. From it the course to Lake Mentasta is nearly due west. On each side of us and converging as we advanced were two tributaries of Tokai River, one of which was reported to head in Lake Mentasta, the other headed to the east and south.

Marshall forwarded this information to me and asked whether the name in the 2008 *Ahtna Place Names Lists*, the Ahtna name **Tak'ae Tl'aa** 'timbered valley headwaters' for an area northwest of Mentasta Lake, might be the origin of Allen's Tokai River.

In the original translation in *The Headwaters People* (Kari 1986:184), the final sentence of Katie John's first segment was:
Mendaesde xona tak'ae yekedeł.
/They would come into the valley at 'shallows lake place' (Old Mentasta).

In 1986 I had treated **tak'ae** not as an Ahtna place name, but as the landscape term that refers to a 'timbered valley' (literally, **tak'ae** means 'water cavity'). Marshall's research prompted me to review names containing the term **tak'ae**. In fact, there is a set of eight Ahtna names in the Mentasta Pass area with the term **tak'ae**, and Katie John in her 1981 narrative actually uses the simplest name, **Tak'ae.**

Thus I recently added **Tak'ae** to the Slana River place name list based on Allen's map, Katie's John's use of the name (4-16) in the 1981 narrative, and Shelly Marshall's sleuthing. While there is no Ahtna stream called **Tak'ae Na',** it seems likely that as Allen and the Mentasta people surveyed names and trails around Mentasta Pass in June 1885, he assigned the name Tokai River (based upon **Tak'ae)** to the stream that heads around Mineral Lake and joins the Little Tok River. Until I read Marshall's article, the association of "Tokai," "Tok," and "Tak'ae" had never occurred to me.

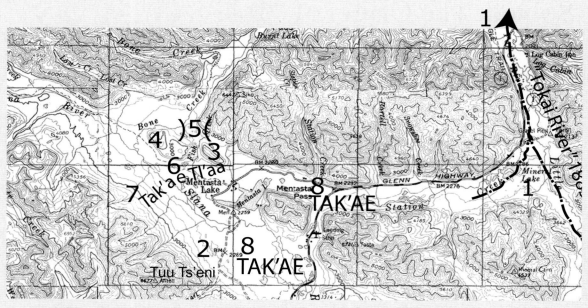

Figure 4-4. Set of eight names with tak'ae **'timbered valley' in the Mentasta area.**

1. Allen's Tokai River   Course of Mineral Creek and Little Tok River
2. **Tak'ez'aann**   Two hills NE of **Tuu Ts'eni** [1-71, mentioned by Jim McKinley and Katie John]
3. **Daniists'en Tak'ez'aann**   West of two hills on Fish Creek
4. **Dadaasts'en Tak'ez'aann**   East of two hills on Fish Creek
5. **Tak'ez'aann Ghatgge 'En**   Pass between two hills on Fish Creek
6. **Tak'ez'aann T'ax Menn'**   Lake in pass between two hills before Bone Creek
7. **Tak'ae Tl'aa**   Northwest area of Mentasta Pass Valley
8. **Tak'ae**   Mentasta Pass Valley, valley bottom north of Indian Pass, and surrounding Mentasta Pass [4-16]

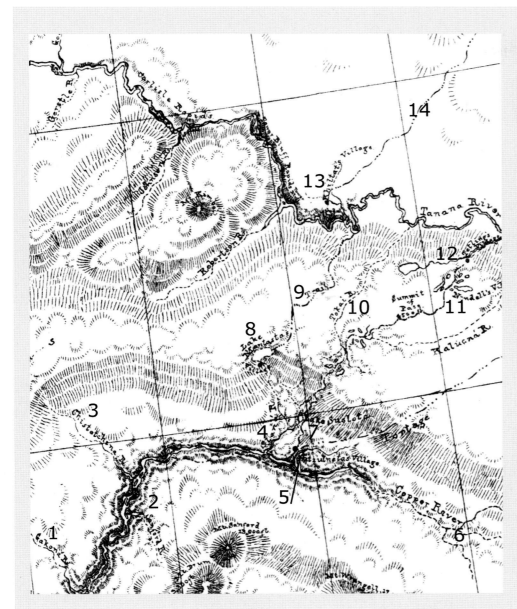

Figure 4-5. Portion of Allen 1887 Map no. 1 at the divide of the Copper and Tanana Rivers.

This a rare and highly important map, the first time these features were mapped and named. Numbered features are:

1 Gakona River, 1-55
2 Sanford River, 5-9
3 Chistochina River, 4-33
4 Slana River, 4-4
5 Batzulnetas, 1-69 (for Chief **Bets'ulnii Ta'**)*
6 Passes and trails to Nabesna R
7 Suslota Lake site, 1-70
8 Mentasta Lake site, 1-72
9 Yerrick Creek Trail to Mansfield
10 Tokai River, Little Tok River (see Figure 4-4)
11 Last Tetlin, Nandell's Village (for Chief **Nandaeł Ta'**)*
12 Tetlin Village
13 Mansfield Village, Khiltat's village (for Chief **K'etl'aadi Ta'**)*
14 Trail to Yukon River noted

* Personal names reported by Fred John Sr.

(2)

4:41
Starting point: Batzulnetas to upper Nabesna River and returning
Ending point: Batzulnetas and New Suslota
Round-trip totals about 80 miles.

Datsii Nataełde kughile'e gha nahwgholnigi.
/I'm going to tell about that lowland place at 'roasted salmon place' [1-69, Batzulnetas].

Nataełde kedi'a' de yet gaa.
/There, where it was called 'roasted salmon place' here.

Koht'aenn 'iinn tseh koht'aenn ye nitezdaetl de tseh Cet'aenn 'ehwdil'aan de.
/Prior to the people arriving there, it was originally discovered by Cet'aenn, 'the tailed ones'.

Cet'aenn hghighaande k'ets'ende, koht'aenn 'iinn gha yet tak'ae kułtsiin dze'
/After they killed the Cet'aenn, the Ahtnas kept that valley (as a home territory) and[2]

łukae gha ye koht'aenn ninidaetl.
/people came there for salmon.
5:10
Ye gaa duu hwghaagha da sii sta' snaan 'iinn 'eł ta yet nihnidaetl
/Recently near there my father and my mother arrived and

dze' ghat yet tah k'enesdzet sii.
/and there is where I grew up, me.

Kadaat kudełdiyede Nataeł Na' kediłende yet du' stsiye Billy yet ghida'.
/Nearby at the next place downriver where 'roasted salmon creek' [4-17, Tanada Ck] flows there my grandfather Billy (Henry) stayed.

Dets'enekaey' 'eł yet hdaghalts'e'. Yet łuk'ae gha gha yet saenn tah hdelts'iix.
/His children stayed there with him. They stayed there during the summer to fish.

Xey tah nahwtak'asi dze', C'et'aan' Hwdittsiic xu tkonii dze'
/In the winter as it turns cold, when it is said 'leaves turn yellow' (late August),

'eł ta 'ungge xona Ts'itu' Tl'aa Ngge' tah k'ena . . . k'enanaełkeldel dze' 'ungge.
/and upland to 'major river headwaters uplands' [4-18, the country at the head of the Copper R] they go back moving nomadically to the uplands.

---

[2] The remarkable story of the 'tailed ones' or 'monkey people' told by Fred and Katie John is in Kari 1986:39–46.

'Unggat tes Sez'aann xe'edił'aan ye t'aax.
/Upland at the hill they named it 'Heart' or 'Inside me', beneath there [4-19, mountain 6580', north of Copper Glacier. See Figure 5-9.]
6:00
Ye t'aax dze' 'ehdelts'iix dze' yihwts'en łuxac'ehdelyiis dze'
/They stayed beneath there and they hunted in groups from there and

debae keghiix gha yet.
/they killed sheep there.

Ye hungge ngge' łuu gha xuh'eł łuhtediił dze' debae yaen' c'a xuh c'ilaen.
/There in the uplands, upland by the glacier (Copper) they would go around and only sheep are there [4-20, Copper Glacier, called Ts'itu' Luu'].

Udzih c'a kol.
/There are no caribou.

Dze' ya'ooxo debae kaen' xu hwgha xey nakedax dze'.
/And they would spend the winter out on that side living on sheep.

Xutah xey hdelts'iix.
/They stayed out in the country in the winter.

'Aeł 'eł kelaax xu.
/They had traps set in the area.

De xona hwtił . . . hwtełgguusi ts'en' ta xona ts' enakedeł dze'
/As it became spring there they would come back out and
6:31
'utsiit yet deghak'ae ninakedeł.
/they would stop again in the lowlands at their home.

Naenn du' kanii hwts'en danggeh Tanaade Menn' xu' kenii.
/As for us, the next place upriver from there, upland they call 'moving water lake' [4-21, Tanada L].

Xuhtah nenenn' ta kughile' dze' kuhtah nin'ta łustediił.
/Within that area was our country and within there we would go around the country.

Nataełde ts'ets'edeł dze'
/We would leave from 'roasted salmon place' [1-69, Batzulnetas] and

'ungge ta Ts'abael K'edigha xu hwgha kets'udeł dze'
/to the upland we would pass by 'by the one with spruce on it' [4-22, hill S of Nataelde] and

'unggat Tanaade gha gha yet ta ts' eneyeł ts'en ye xanggat.
/upland there at 'by moving water place' [4-23, Tanada L outlet site] there, we would camp at the next place upland.

Men Diłende yet ta xona yet cu nits'edeł.
/From there we would stop at 'where it flows into lake' [4-24, Camp Ck into SE shore of Tanada L].

Yet ts'e yihwts'en łu . . . łuhtedeł dze' debae kol dze'.
/Then they would go out from there and if there were no sheep.
7:09
Yihwts'en ts'ents'edełi 'eł xona Łedidlende ta ye nits'edeł.
/We started out again from there and then we stopped at 'where streams join' [4-25, Goat Ck].

Łedidlende yet ta xona debae keghiix dze'. Yii debae 'eł yet sdelts'iix.
/They killed sheep there then at 'where streams join'. We stayed there with that sheep.

Yihwts'en xona danggeh ta c'ena' ngge' ta kets'edeł dze' yi ya xungge' kecdilaa de.
/From there then in the uplands, as we go on upland, and there are names in the upland there.
7:31
Ts'akae . . . Ts'akae Ggan Nats'iłbaał keniide.
/Where they call 'the thin lady was lowered on a rope'[3] [4-26, creek off Tanada Peak into Goat Ck].

Ye xuh kets' udeł dze' ye xanggat da men, Men Niłgha'aa Delyaade.
/They pass by there and the lakes up there, 'lakes connected together' [4-27, lakes at summit of Goat and Jacksina creeks].

Gha ye xu'enn' cu kats'edeł dze'.
/There we go up over to the other side also (to the Nabesna R drainage).

Ye xu'en xona Tsi . . . Tsic'ełggodi Tl'aa keniide ye kets'edeł.
/There over to *f.s.* 'rock is chipped headwaters' [4-28, Jacksina Ck headwaters] we came to there.

Tsic'ełggodi keniide 'adii du' Jacksina Nondlae 'iinn Jacksina kehdił'aan.
/Where they now call 'rock is chipped' the white people have named the place called Jacksina.

Jack . . . gha yet Tsic'ełggodi Na' tsen ts'ets'edeł.
/Jack . . . there at 'rock is chipped creek' we come out to the lowlands [4-29, Jacksina Ck].

---

[3] Katie tells the story of this place name: "They killed one sheep up on the mountain, and it fell down into the steep canyon. They put a rope around this woman and lowered her down. She cut up the sheep, and they raised the meat up, piece by piece. Then they lifted her out."

Yets'e 'utsiidze yet Ts'iłten' ka . . . Kats'etses Na' keniide
/From there from lowland (the Nabesna R) to where *f.s.* they call 'we lift up a bow creek'[4] [4-30, Pass Ck].
8:08
Yet xu'aadze ta kanats'edeł dze'.
/We come back up from the other side.

Ye datsendze stedeł ta xona Ts'es T'aax keniide yet cu nits'edeł dze'.
/When we come from the lowlands, then we stop at where they call it 'beneath the rock' [4-31, Wait Ck].

Ye cu debae gha sdalts' iix.
/There, too, we would stay for sheep.
8:18
Gha ye ts'en niłkenastedeł dze' xona 'u'aadze łe . . . Łedidlen de ninats' edeł.
/From there we would make a round trip and from the other side we stop again at 'where streams join' [4-25, Goat Ck].

Gha ye Łedidlen de yedu' xona c'etsen' dak'ae kukuł'aen de.
/There at 'where streams join' then they kept a meat cache.
8:31
Ye tah yet du' xona all niłc'aa kulaen dze' łuhtediił dze'.
/There they would go around (hunting) for various (foods) available.

Debae keghiix.
/They would kill sheep.

Ye all c'etsen' niłyikelaes dze' yet xona nixii'eł nadetseh naltiin dze'.
/There they gathered together all that meat, and there then they would relay it on ahead.

Tanaade Menn' tene keyelyiis.
/They would bring it along the trail to 'moving water lake' [4-21, Tanada L].

Men Diłende ye cu ninats'enedeł dze'.
/We would camp there again at 'where it flows into lake' [4-24, Camp Ck on SE shore of Tanada L].

Tanaade yet ts'en Ts'abael K'edigha cu nints'enedeł.
/From 'moving water place' we camped again at 'by the one with spruce on it' [4-23, 4-22].
9:00
Cu Nataełde K'enats'edeł.
/Again we got back to 'roasted salmon place' [1-69, Batzulnetas].

---

[4] Katie says that this name describes a very steep place in the mountains where one man would go up and then the next one would hand up his bow so that he could be pulled up.

Figure 4-6. Molly Galbreath's painting of New Suslota village or Bes Ce'e **[4-1] as it looked in the 1940s.**

The cabins, left to right, belonged to Sanford Charley, Guy John, Frank Sanford, Alec John, and Shorty Frank.

Nataełde yi ta xona nints'edeł dze' ye sdelts'iix.
/When we returned to 'roasted salmon place' we stayed there.

Xona nanin'detiisi
/Then the ground would be freezing again

'eł 'utgge' Bes Ce'e dze' ţgge' ta xona nastedeł dze'
/and then above to 'big bank' [4-1, New Suslota] up above we returned and

'utgga Bes Ce'e ta kanats'edeł dze' yet cu nekonagh' 'eł kuzdlaa dze'.
/we went back up above to 'big bank' and we also had houses there.

Yet xona xey nats'edax.
/Then we would spend the winter there.

Yihwts'en xona 'aeł 'eł sta' 'eł snaan 'eł 'aeł 'eł ta xey kelaax hwna,
/From there then my father and mother while they set traps during the winter,
9:32
xey xu'eł nakozet.
/the winter would pass by for them.

Yet ts'en 'aeł yunyeggaay ta all niłtahdze' c'altsiił dze' tsa' zes niłyihdelaes dze'.
/From the traps were foxes of all color phases and they gathered together fur.

(3)

9:42
Starting point: Slana area
Ending point: areas to south toward Mount Sanford
Route totals about 35 miles.

Xona yi 'eł ta xona da . . . datsen Stl'aa Caegge dze' tsene ta xii'eł tadeł dze'.
/Then there to lowlands at 'mouth of rear river' [1-68, mouth of Slana R] they went with that to the lowlands.

'Utsii Stl'aa Caegge hneyeł dze'
/They would camp downland by 'mouth of rear river' and

yihwts'en xudaa' Ts'itu' K'et daa' htediił dze'
/from there they would go downriver to 'on major river' [4-32, camp near mouth of Slana R on Copper R] and

'daa' Tsiis Tl'edze' Na' yetah xona tsa' zes 'en 'ehdeliis.
/downstream at 'blue ochre river' they would sell the fur [4-33, Chistochina R and village].
10:05
Ye xu tene xey tene Ts'itu' K'et daa' xey tene kughile'.
/There the trail, the winter trail, went downriver from 'on major river' [4-32, Copper R].

Ye gga' Tsiis Tl'edze' Na' yet du' xona tsa' zes 'udetkaetde yet Nondlae zdaa de.
/Then there at 'blue ochre river' [4-33, Chistochina R], there the fur was purchased there where a white man stayed.

Yetah 'en kahwdelaesi c'aan kunesi 'eł kanakedeł.
/When they sold it, they obtained food and came back up,

Xona.
/That's all.
10:30
Gaa Stl'aa Caegge kedi'aan de.
/Here is what is called 'mouth of rear river' [1-68, Slana].

Datggat tes ggaay nez'aan c'a Dzii Koley xii'edił'aann.
/That little hill up there they named 'the deaf one' [4-34, hill at Slana].

Yii c'a xiigha nakalniisi.
/They told this about it.

Denae udzii kughistle'e.
/A man was deaf.

Yen deyaagge, yen uk'et ni'ilniic xiiłnii.
/When he died he was buried on it (hill), they say.

Yii gha' Dzii Koley xu'e xii'edił'aan.
/That is why they named the place 'the deaf one'.

Gha yii tes ggaay yii c'a ye t'aax dze' gaa nisiltaen de.
/Below that little hill here is where I was born.

Gha ye nekonagh' kughile'de.
/Our house was there.

Danooxu kughi'a'
/It stood a ways upriver.

Gaa duu nondlae 'Unsogho ts'en kezyaann.
/Here a white man came from 'distant area in front' [4-35, "Outside", distant places outside of Alaska, the common Ahtna name].

Yen sta' ikonagh' sta' konagh' kughikaet dze' yi t'aax daninaa dze'
/He bought my father's house and he moved in below there and

xuk'a 'en c'aan 'eł 'en telaesi gha nikuni'aan dze' store nikuni'aan.
/he started selling food and he built a structure and he built a store.

Yi gha'aat du' sta' konax nakułtsiinde.
/Off from there there my father built another house.

Yi t'aax c'a nisiltaen de.
/In there is where I was born.

Datsuux yae' Ts'itu' K'et keniide Ts'itu' ts'idiniłen.
/Down below there, where they call it 'on major river' flows out [4-32, camp on Copper R below Slana R mouth; 1-6 Copper R].

This is Copper River, close by Copper River.

Stl'aa Caegge da'aa niit kediłen.
/'Mouth of rear river' [1-68, mouth of Slana R] flows in there to the upstream.

Gaa du' xona Stl'aa Caegge kedi'a' de.
/Here then, it is named 'mouth of rear river' [1-68].
(break)
11:50
Danuux yae' Ts'itu' danaasts'en yae',
/In the upstream there on the other side of 'major river' [1-60, Copper R],

ts'elk'ey tes ggaay nez'aani Lts'uusi xii'edił'aan.
/one little hill sticking up that they named 'pointed one' [4-36, hill on south bank of Copper R above Indian R].

'Udaat ighadaat Tsedghaazi Na' ghaaghe nez'aann du' Nexk'aedi xii'edił'aann.
/Downriver, down from it 'rough rock creek' [4-37 Boulder Ck], nearby a hill stands that they named 'lookout' [4-38, point on Capital Mt].

Nexk'aedi.
/'Lookout'.

Xiik'et kiike'ideł dze' xiik'et ts'enaexdelts'iix nkohnesi 'uka.
/They would go up onto it and they could sit on it looking around for game.

Yii gha' c'a Nexk'aedi xii'edił'aan.
/This is why they call it 'Lookout'.

Nanuuxu yae' K'ełt'aenn t'aax yae'.
/Upriver that way, it is beneath 'K'ełt'aenn' [4-39, Mount Sanford, also called Hwniindi K'ełt'aenn].
Ends 12:41

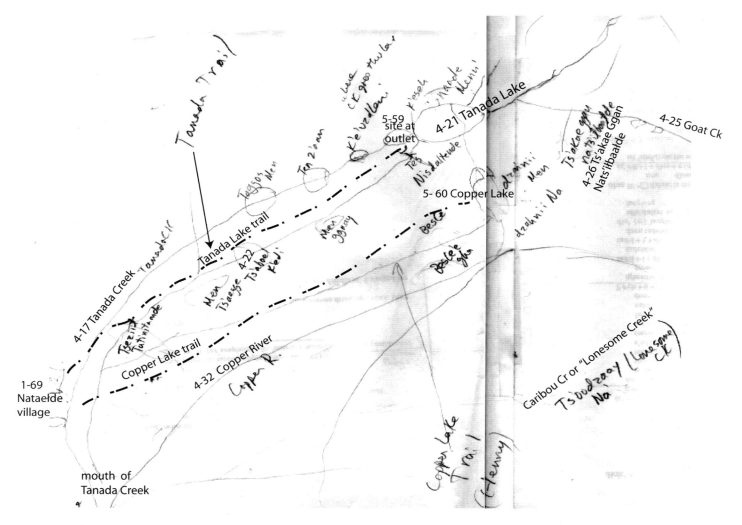

Figure 4-7. Sketch map of Copper Lake, Tanada Lake, and the Goat Creek area by Katie John made on May 30, 1984. She mentioned eight names that are referred to in Chapters 4 and 5 by Katie or by Adam Sanford as well as about 10 other Ahtna place names and many other features.

5

C'uka Ts'ul'aen'i gha Nen' Ta'stedeł dze'

How We Went Hunting Out in the Country

Figure 5-1. Portrait of Adam Sanford in about 1979.

Photo by Linda Weld.

Adam Sanford

Adam Sanford, the younger brother of Sanford Charley, was born in 1887 and he died in 1983 at age 96. Between 1974 and 1981 I worked with Adam on five occasions. The following narrative was recorded in one session on June 13, 1981 with Kate Sanford and Katie John in attendance. This appears to be a rather comprehensive summary of the seasonal movements of the Chistochina Ahtna band. Adam's athleticism and intimate firsthand knowledge of the country are evident throughout.

Previously published in Kari 1986:161–180, however, this version follows the original audio more rigorously, and there are many edits to the place names and directional terms. Also added to the end of Part 1 is a 30-second segment from the only other time I tape-recorded Adam, on September 26, 1981. He happened to mention **'Uzdledi T'aax** [5-22; see Figure 5.3], a ridge and lake district probably on the east side of the Sanford River, that was not in the 2008 version of the Ahtna place names. This one brief mention of a place name underscores the importance of maintaining both lists of place names as well as texts about places.

Part 1 is probably the most detailed Alaska Native language account of travel and resource procurement in proximity to glaciers and high mountains. See the discussion at Figures 5-3 and 5-4. In the 1983 *Ahtna Place Names Lists* and in the 1986 version we had not known the exact locations of the named side streams of the Sanford River.

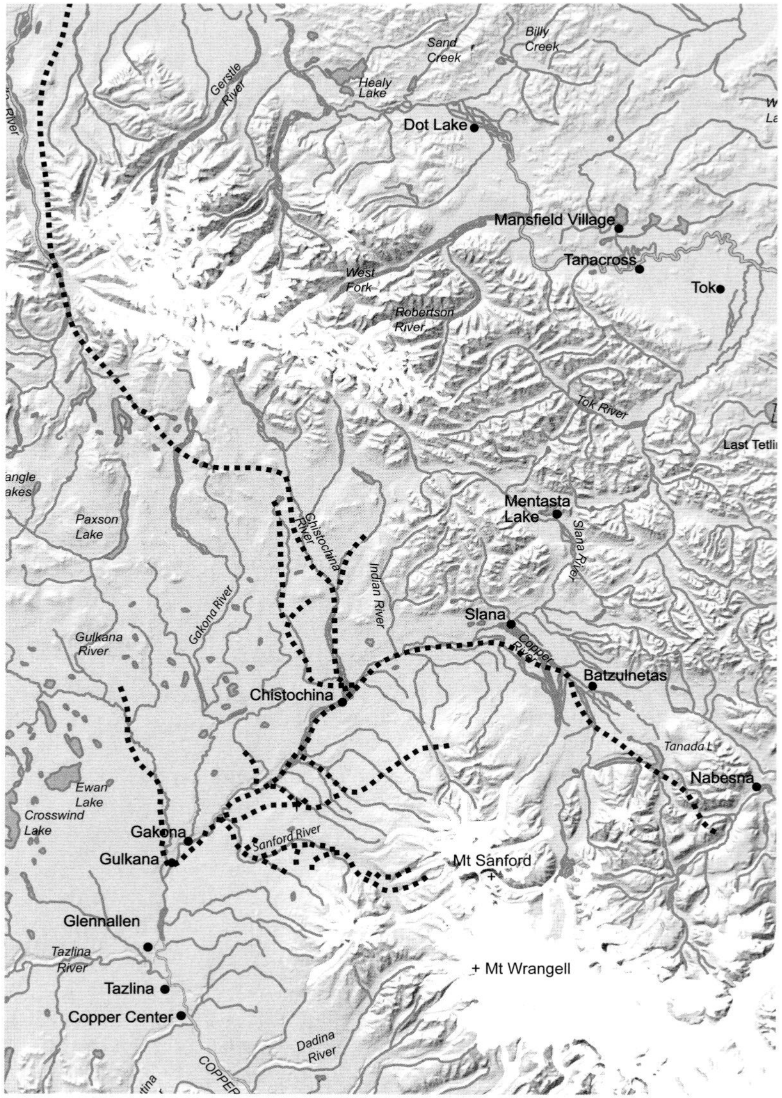

Figure 5-2. Routes traced in the Adam Sanford narrative. Part 1 covers about 110 miles; Part 2 is about 50 miles; Part 3 covers about 160 miles (one way); and Part 5 covers 160 miles (in two routes).

(1)

Total: 29:08. Sound file chp5-adamsanford.wav.
Starting point: mouth of Sanford River, to upper Sanford River, back to Copper River, upstream to Boulder Creek, to Chistochina
Ending point: Chistochina area
The routes total about 110 miles.

Ts'itael Caegge xak'a su sc'ilaen. Yi yet su nisiltaen.
/I was born right at 'mouth of river that flows straight' [5-1, the mouth of the Sanford R]. There I was born.

Yet su xa'atle k'enesdzet de su duugh nats'idaetl, you know.
/Mainly I grew up there and we would come back there.

Ts'itael Caegge, Ts'itael Caegge nanaa neghak'ae.
/Across from 'mouth of flows straight', 'mouth of flows straight' (was) our home.

Stsucde 'iin ye hdaghalts'e'.
/My grandmothers lived there.

Stsucde 'iin ye hdaghalts'e'.
/My grandmothers lived there.

Yet su stsucde 'iin gha nats'idaetl.
/We would go back there to my grandmother's people.

Yihwts'entah xona Ts'itael Na' Ngge' nen' tah 'skultsiin.
/From there we had hunting territory up 'river that flows straight uplands' [5-2, Sanford R uplands].
:35
Łuk'ae gha 'sdelts'iix Ts'itael Caegge.
/We would stay for salmon at 'mouth of flows straight' [5-1].

Łuk'ae nits'ełcet.
/We would put up fish.

Xona xa'atle hwdezeltiy.
/Then I was just almost a teenager.

Yihwts'en ta xona nen' ta'stedeł debae ka deniigi 'eł udzih 'eł.
/From there then we would go up into the country for sheep and moose and caribou.

Ts'itaeł Na' Ngge' 'stedeł dze' 'unggat Ba'stadełi hwts'e' ta 'sneyeł.
/We would go to 'river that flows straight uplands' [5-2], and we would camp up at 'the one we go out to' [5-3, hill north of Sanford R mouth].
1:03
Duu yihwts'en xona 'unggat Natii Caegge yedu' xona yedu' xona nits'edeł.
/From here, then over to 'natii mouth' [5-4, creek into Sanford R from south], then we would stop there.

Yihwts'en xona Natii Na' Ngge', xona 'utgge yii xungge' tah kudełdiye.
/From there, then in 'natii river uplands' [5-5], then above there to the uplands is a short distance.

About yet nduugh miles gha kulaen?
/How many miles is it to there?

Seven (or) eight miles, I guess.

Yet su xona debae ka 'stedeł.
/There we went for sheep.

Debae ts'eghaax gha yak'a.
/We would kill sheep right there.

Yii kaen' taade yi yet 'sneyeł.
/We stayed there three days (living) on it.
1:35
Du' yihwts'en ts'inats'edeł dze' 'unggeh.
/From there, then we would start out again to uplands.

'Utggu daagha ngge' ngga Ts'itaeł Tl'aa hwts'e',
/Up above the tree-line upland to 'headwaters of river that flows straight' [5-6, head of Sanford R],

yihwts'en 'unggat Tsaani 'Aeł Na' yet kats'edeł.
/from there on upland we reached 'bear trap creek' [5-7, creek from south above Natii R].

Yet kanaa debae una' c'ilaen, you know.
/Across from there, there are sheep on that creek, you know.

Yet cu debae ka łu'stedeł. Debae ts'eghaax, you know.
/There we would hunt again for sheep. We would kill sheep.

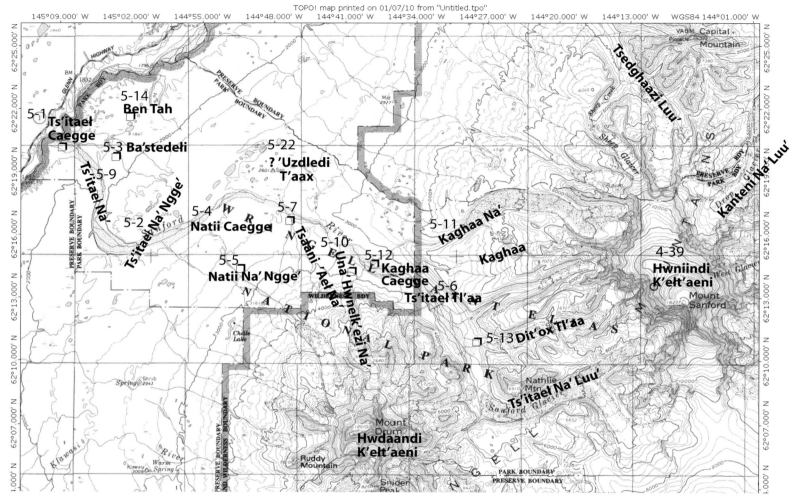

Figure 5-3. The first four minutes of Part 1 can be analyzed in several ways. This is one of the finest-grained accounts ever recorded of Athabascan band movements in glaciated mountain country. Marked on a U.S. Geological Survey map of the Sanford River with reference numbers are 14 place names Adam Sanford mentions in Part 1 as well as several names that he does not mention. The distance in a straight course from the mouth of the Sanford River to the uppermost destination, **Dit'ox Tl'aa** [5-13], is about 37 miles. The upper Sanford River plain seems to be over 2,000 feet in elevation. The ridges and domes where sheep were hunted must be 2,500 to 4,000 feet in elevation. In the first four minutes of Part 1, Adam uses the verb **d+l+ts'ii** 'plural stay' seven times and the verb **n+yeł** 'spend the night' four times. Approximate locations for some of these overnight camps are noted with the symbol ☐.

Ye naxaełts'eldeli kae taade nk'e yet ts'eneyeł.
/With what we were packing back, we would camp there three days.

Duu yihwts'en xona Natii Na', Ts'itaeł Na' kenats'edeł dze'.
/From there we would go back to 'natii river' [5-8] and to 'river that flows straight' [5-9, Sanford R].
2:08
Ts'itaeł Na' Ngge' 'ungge na'aa ngge' 'stedeł dze'.
/In the 'river that flows straight uplands' [5-2] we go upland out and over in the uplands.

'Unggat Una' Hwnelk'ezi Na' ye nits'edeł.
/Upland we would stop there at 'its creek is a brown area' [5-10, creek into Sanford R to the south above Natii R].

Yedu' xona debae c'ilaen, you know.
/There were sheep there.

'Unggat łuu gha yet k'a 'sdelts'iix.
/At an upland place there we stayed right by the glacier.

Nduude k'e yet 'sneyeł.
/We camped there awhile.

Tiye yet 'sdelts'iix hwna debae kae.
/We stayed there a long time while (living) on sheep.
2:37
Ts'itaeł Na' łuu ts'e' kudełdiye yehwts'en.
/From 'river that flows straight' to the glaciers [first glaciers to N on Sanford R, 5-9] are nearby.

Oh, about three miles, two miles, I guess.

Yet xona uyii naane skets'enaes.
/Then we would move across and into it (canyon) there.

'Unaats'en cu debae gha yet 'sdelts'iix.
/On the other side there we stayed for sheep.

Duu yihwts'en xona ts'inats'edeł dze' 'unggase'.
/From there then we would start out again down from the uplands.

Danii ts'en nggase' tat'aa nggase' 'adii na'stetnaes.
/From upriver, from the uplands to the valley from the uplands now we would return.

'Ungga yet su ts'edelts'iix tah Kaghaa Na' hwdi'aande su.
/We stayed upland there at a place called 'along expanse creek' [5-11, creek into Sanford R from north].
3:14
Kaghaa Caegge cu ts'edalts'iix.
/We stayed also at 'along expanse mouth' [5-12].

Yehwts'en ts'inats'edeł dze' 'utsii 'utsiit gge' kats'enaes, you know.
/Starting out again from there, in the lowlands, in the lowlands and upward we would move up above.

Dit'ox Tl'aa ts'e', Dit'ox Tl'aa ts'e' kats'enaes, you know.
/To 'nest headwaters', we would go move to 'nest headwaters' [5-13, creek from north off Mount Sanford].

Big glacier (is) there, you know.

Dit'ox Tl'aa du' yii cu ts'edelts'iix debae gha.
/We stayed there at 'nest headwaters' by the sheep.

Sesyaan' sesyaan' yaen' una' c'ilaen.
/There are rams, only rams on that creek.

Yii ts'eghaax xona yii kae yet 'sdalts'iix.
/We killed some and we stayed there on that.
3:40
K'a xona yet hwts'en xona na'stetnaesi, ohh dahwtnełdak.
/Then as we moved back from there, oh it (the canyon) was steep.

Niłk'aedze' dahwtnełdak xona, saane tah kats'enaes.
/It was steep on both sides and then we could barely move up.

Xona dae' ts'en ben, Ben Tah keniide yet c'a yii nits'enaes.[1]
/Then on that side there, a lake, where they call 'among the lakes' [5-14, lakes north of Sanford R], there we stopped.

Xona 'utsuughe ts'abaeli nae' nel'aa xu xu xu ninats'etnaes, xona deniigi ka.
/Then we came to the lowland area where the tree line extends upstream (direction toward upper Copper R), and we stopped again, then for moose.
4:22
Xona yehwts'en łu' xona deniigi ka dinats'enaes. Ts'ełk'ey deniigi ts'ezełghaen.
/Then from there we went in again for moose. We killed one moose.

---

[1] There is an abrupt change in location as Adam Sanford starts a new tour from 5-14, the lakes near the Copper River (see Figure 5-3).

Not much deniigi c'ulae'i, that time.
/There were not many moose (around) at that time.

Duutlek'e ya yuughe niłdenta yaen'. Udzih yaen' c'ilaen, you know.
/Only here and there occasionally. There were only caribou, you know.

Udzih yaen' c'ghile'.
/There were only caribou.

Yehwts'en nadaeggi deniigi one k'a ts'ezełghaes, ts'ełk'ey yehw.
/There we could kill only two or one moose there.

Du' yehwts'en ye nae' 'stenae[s] 'utggu daaghe nae'k'e.
/From there we started out upstream to above the timberline on the upstream.

Kaniit Una' Tuu Koley Na' yehwk'e nats'etnaes.
/The next place upriver we went back to 'its creek has no water-creek' [5-15, creek from east above Sdzedi Na'] .
5:03
Ye tah xona xona udzih udzih tana'stedeł.
/There then, then caribou, we came back among caribou.

Da'ooxe teye nelt'e'i nsiil laex.
/Out from there, there were lots of caribou, summer caribou 'warm ones'.[2]

[JK speaks]
Q: This is on the Mt. Sanford side?

Una' Tuu Koley Na', you know.
/'Its creek has no water-creek'.

No water, that creek. Big river, big creek, though.

Duu yet hwts'en xona 'uniit Saas Dzeł k'et ninats'etnaes.
/From there, then we would stop again upriver at 'sand mountain' [5-16, mountain on south side of Boulder Ck].

'Utggat tes ghak'aay yi ninats'etnaes.
/Up on the flank of a hill we would stop.

---

[2] **Nsiil** 'the warm ones' are caribou that stay up in the mountains during warm weather to avoid droves of mosquitoes.

Figure 5-4. Mount Sanford. The Sanford River heads in glaciers off the Wrangell ice field between 12,010-foot Mount Drum and 16,237-foot Mount Sanford. Mount Sanford was photographed on April 11, 1982, after a landslide in which debris fell about 10,000 vertical feet. In the foreground is a bend in Sanford Glacier. Immediately below the frame of the picture is **Dit'ox Tl'aa** [5-13], the farthest upstream drainage mentioned by Adam Sanford. This stream and glacier are three to four miles from the great vertical face of the mountain.

Photo by University of Alaska Fairbanks, Geophysical Institute, courtesy of Carl Benson.

Tseles tseles gha yet nits'enaes.
/We stopped there for ground squirrels.
5:38
Snaan 'iin łuu ts'e' tseles gha kac'etl'uun.
/My mother and them toward the glacier set snares for ground squirrels.

Sii yet sii du' k'alii tseles gha hwdeł'aha.
/As for me, I didn't care much about ground squirrels.

Yet yak'a natesdaas dze' du' yet tseles gha teye 'sneyeł.
/I would just walk around there, and so we camped there a long time for ground squirrels.

Teye hwnelt'aede 'sneyeł.
/We camped quite awhile.

Tseles keghaax.
/They killed ground squirrels.

Kiinansilii' ts'e' dakiidghilaes.
/They would skin them and hang them up.

Kiiłggan.
/They dried them.

Du' yeghik'ey yeł yegheł t'adezdlaa.
/There she covered it (meat) with birch bark and she put that in the dog packs.

Yi kiinelt'os dze' xona cu ts'inats'etnaes.
/They stuffed it in (the meat into the dog packs) and then we started off again.
6:12
Du' xona Tsidghaazi Tl'aa nanaa 'uniit,
/Then to 'rough rock headwaters' [5-17, upper Boulder Ck headwaters] across there and upriver,

fish camp nanaa 'uniit Kediłeni Na',
/the fish camp (Adam's current fish camp) across and upriver at 'creek that current flows in' [5-18, creek into Copper R south of Boulder Ck],

'utggat utl'aat Tsidghaazi Tl'aa yet yehwk'e nats'etnaes.
/we would return up its headwaters, to 'rough rock headwaters' [5-17].

Yii cu debae gha. Yet cu debae una' c'ilaen.
/There also is for sheep. That, too, is a sheep stream.

Gha Debae gha yet cu 'sdelts'iix. Teye 'sdelts'iix yet, debae ts'eldaełi kae.
/We stayed there too for sheep. We stayed there a long time eating sheep.

Yet su yehwts'en Kateni Na' Ngge' 'eł Kateni Na' Ngge' cu 'stedeł.
/From there we went to 'ascending trail creek uplands' [5-19, creek W of upper Drop Ck], to 'ascending trail creek uplands'.

Yet yii c'a big river, big creek.
/That, too, is a big river, big creek.

Ye naak'e łu'stediił xu debae gha.
/We would go around on the mineral licks for sheep.

Kudełdiye dze' debae c'ilaen, you know.
/Sheep are nearby, you know.
7:00
Kudełdiyede k'a ts'eghaax.
/We could kill them right nearby.

Naxaełts'eldiiłi dze'.
/We packed them back.

Ukae k'a ts'edelts'iix.
/We just lived on that.

Udzih cu 'utggu utak'a.
/Caribou also up above were interspersed.

Udzih cu utak'a c'ilaen, you know.
/Caribou were also interspersed, you know.

Yii k'a ts'eghaax, you know.
/Those we would kill.

Uk'e 'sc'eyaani kae na'stediił hwnah.
/While we would go about living on what we ate.

Hn nekokaedi gaa k'alii sut'e' c'a kestlaeghe hwna.
/Well, our food did not run out here.

Cu ka'aat su debae gha ts'ets'edeł.
/Again we would come out to the next place over for sheep.

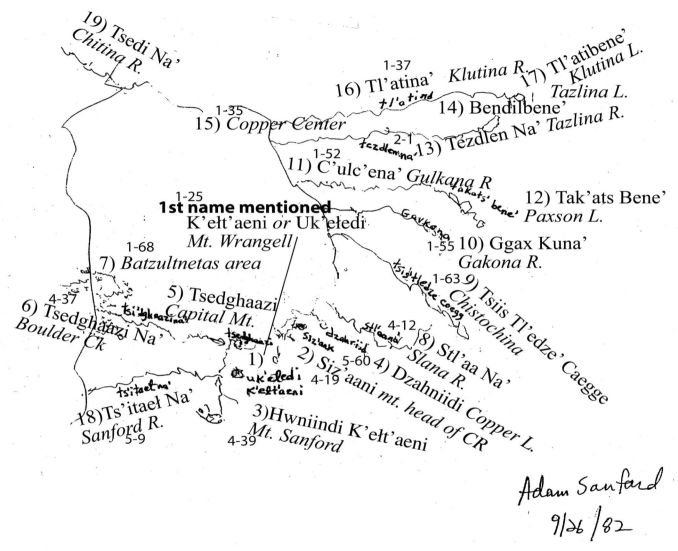

Figure 5-5. The Wrangell Mountains and the Copper River drainages sketched by Adam Sanford on September 26, 1982, in Chistochina. My tape recorder was on as he sketched this in about five minutes. Numbers with) are in the order that Adam drew and named about 20 places. Fifteen place name reference numbers are noted as well. Adam's sense of the regional geography is fascinating. The region was perfectly coherent in his mind, although he transposed some features due to his limited eyesight.

Yihwts'en xona Tsidghaazi Na' tggase' na'stetnaes.
/From there we move back from above at 'rough rock creek' [4-37, Boulder Ck].

Nanaat nanuughe Snuu Caek'e ts'inats'etnaes.
/We came out across in upriver area of 'brushy mouth' [5-20, Sinona Ck mouth].
7:38
Yet Snuu Caegge ghat yet Snuu Caegge gha 'sdalts'iix hwna.
/By 'brushy mouth', then by 'brushy mouth' we would stay.

Yeł łuk'ae 'eł dzaxdze' dzenax 'eł c'ilaen, you know. 'Sc'eł'iix.
/There were salmon and lots of fermented fish, you know. We would make this.

Snaan dzenax c'eł'iix.
/My mother would make fermented fish.
8:00
Yii kae xona ne'eł ten ghan c'ełaex xona.
/Then with that (food) with us the shelf ice (on the edge of the river) would appear.

Katie John asks:
'Unoo 'uniit gha yet da Ts'itu' K'et kedi'aan?
/In upriver area, at a place upriver there, there it is named 'on the major river' [5-21, fish camp near Chistochina]?

Adam Sanford:
Aanhaan.
/Yes.

Xona yet tseh tah hdaghalts'e all netseh tidakestlaak.
/Those the people who lived there previously all had died before us.

Kutl'ahwdalnen.
/They all died off.

Ts'inst'e' ye ghida'en yaen' gha kats'ezdaetl, you know.
/Only an old lady who used to live there, we came up to her there.

Gha yet su xona hwt'ae' kats'ezdaetl.
/There then we all just came up there.

Tsiis Tl'edzi Caegge.
/To 'mouth of ochre river' [1-63, Chistochina village].

Tseh su 'udaa' Ts'itaeł Caegge neghak'ae kughile'.
/Previously our home was down river at 'mouth of flows straight' [5-1, Sanford R mouth].

Yihwts'en 1906 du' Tsiis Tl'edze' Caegge kats'ezdaetl.
/From then we came to 'blue ochre mouth' [1-63, Chistochina R mouth] in 1906.

Yihwts'en k'ali'i 'udaa'a nats'idełe.
/After that we did not go back downriver.

Yihwts'en gha łu nanaa 'uniit
/From there (it was) across and upriver (from the Sanford River site).
Break
8:49
Inserted excerpt from 9/26/81, Adam Sanford, AT5029 or AT49
'Uzdledi T'aax dze' ugheli ye diłt'e'.
/'Beneath the one that melts' is a good place there [5-22, possibly ridge 2601' on N side of Sanford R. See Figure 5-3].

Uluu' ta kol.
/It has no glaciers.

Ben yaen' tsaas uk'e t'ahdiłt'eye duugh.
/There are just lakes (there) and the use it (to obtain) Indian potatoes there.

About foot of water, that's all. No glacier. Nothing. Just cover there.

Gaa Hwniindi K'ełtaeni t'aax ts'idiniłen.
/Here from below 'upstream k'ełt'aeni' [4-39, Mt Sanford] the current flows out.

Yi t'aax su ugheli diłt'ae
/Below there is a good area.

Saen ta ben delaa xuk'a, ukaziyaade łdu' 'utggadi nateltsox.
/In the summer there are lakes. When you go up there, above there it is turning yellow (color).

Yae' ba'steltsełi xuk'a.
/We would run that way toward it.

Ugheldze' 'unaan' uk'e łketayaał
/He can go across upon it nicely.
9:24 insert of 9/26/81 ends

(2)

Starting point: Chistochina, up Chistochina River and back
Ending point: Chistochina area
The route covers about 50 miles.

Xona nahwluude kołaex.
/Then when it became fall.

Xona nahwluude kołaeghe hwna,
/then when it became fall,

xona c'uka ts'ul'aen'i gha nen' ta'stedeł, nen' ten k'edze gha.
/then we would go hunting out in the country on the frozen ground.
9:38
Tsiis Tl'edze' Na' Ngge', Tsiis Tl'edze' Na' Ngge' 'stedeł dze'
/'Blue ochre river uplands' [5-23, Chistochina R Uplands], we went up to the 'blue ochre river uplands' and

nanaa 'unggat Katl'abese' gha yet 'sneyeł.
/we would camp across and upland of 'rear riverbank' [5-24, hill on W side of Chistochina R, three or four miles above the mouth].

Cu yihwts'en 'unggat 'ungga 'Uzdledi T'aax dze' ye nits'enaes de.
/Again, upland from there, upland from 'beneath the one that melts' [5-25, hill 3340' west of Chistochina R], there we stopped.[3]

Saen tah hwna tsuugi ka ni'aełts'elaes, you know.
/During the summer we set traps for marten.

Yihwts'en xona tsuugi ts'eghaax xu.
/From there we killed marten.

K'alii sut'e'i tsuugi ts'esghaaghe.
/We did not catch fine marten.
10:13
Yihwts'en xona ts'inats'etnaesi na'aat.
/From there we went out again out away.

Dangge Snuu Na' 'ungga Snuu Tl'aa yet na'aa ts'etnaes.
/Upland from 'brushy creek' [5-26, Sinona Ck], we would move beyond to 'brushy headwaters' [5-27, head of Sinona Ck].

---

[3] This place [5-25] has the same name as the recently recognized name [5-22, possibly ridge 2601' on N side of Sanford R]. With the consolidation of place names by drainage, it is clear the ridge 5-22 is a distinct feature.

Yihwts'en Snuu Na' nggase' t'i ts'inats'edeł.
/Then we would come back down from the uplands of 'brushy creek' [5-26].

'Unggat 'Uzdledi T'aax dze' yet yet cuu tsuugi gha 'sdalts'iix.
/We would stay also for marten upland of 'beneath the one that melts' [5-25].

Yihwts'en ngga' Snuu Na' ts'inats'edeł dze' yet Snuu Caegge nanaa 'unggat.
/From there we would come back out of uplands of 'brushy creek' [5-26, Sinona Ck] to its 'brushy mouth' [5-21] across and in the upland.

Tsiis Tl'edze' Caegge yet ts'inats'edeł.
/We would come out again from 'blue ochre mouth' [1-63].

Yihwts'en nggase' naat nggase' ts'ints'edeł nanaa 'unii.
/From there we come back out from the uplands across and from the uplands, and across and upriver.

Neghak'ae yet nints'edeł.
/We stopped there at our home.
10:54
Du' yihwts'en xona xona nunyeggaay gha cu ni'aełts'elaes nanaats'en.
/Then from there, then we set traps for fox on the across side (the river).

Nanggat Ketne'aay,
/Upland at 'the one standing by itself' [5-27, hill 2531' south of Chistochina],

down where you see that hill, Ketne'aay.
/'the one standing by itself'.

Ye daniits'en tes tah 'utgget nunyeggaay kaen' kadest'aan xu tkut'ae.
/There on the upriver side up among the hills there are many fox dens.

Du' yii xutah 'aeł ts'elaax nunyeggaay gha.
/Right there among them (dens) we set traps for fox.

Duu yet hwts'en xey xuk'a 'aeł ts'elaax, you know.
/From there we had traps set all winter, you know.
11:23
Xuk'a nunyeggaay ts'eghaax uyii da ukol gha.
/We kept killing foxes, there was no lack of them (being so many of them).

Dzaxdze' nunyeggaay c'eghile', yii. Ggax c'a su ghile'.
/There were so many fox, gee. There were rabbits, too.

Yii łek'a k'alii 'unse' t'ilaale. Cross fox dollar-and-half dighile'.
/They (skins) were not worth much. A cross fox was (worth) one-fifty ($1.50).
11:55
Red fox (was worth) one dollar. Lynx (was worth) two-and-a-half ($2.50), big one.

Niduuy du', niduuy du' two-and-a-half dighile'i.
/Lynx then was $2.50.

Yedi k'alii sut'e'i c'a udicaaxe' c'ilaele. Belzaaze' t'ae' naghaltsaek'e.
/However, no one had much money. The money was really scarce.

Yihwts'en su xona store gaa nikults'et.
/Then a store was established here.

Naat Nanii Ghaay [5-28, store location on S bank of Copper R].
/Across there 'along the upstream place'.
12:09
[At] Cut Off, you know.

Cu yełu', Tsiis Tl'edze' Caegge tah ts'e' tene dunae' de.
/Right there where the trail goes upriver from 'blue ochre mouth' [1-63, mouth of Chistochina R].

Yet xona niłghaay gha nikults'et. Store negha nikults'et.
/There then a store got established. We got a store.

Yet xona k'adii hwghaaghe xona kughile' sii tey c'a łdadenelnen de.
/That was not that long ago, after I had been alive a long time.

Du' yii yii 'eł na'sdelyiis dae' desnen' koley.
/Then we would give him that (in exchange) fur.

Gha yuugh hwdetaey' tah nic'ootket tah.
/Things there were just purchased with credit.

C'aan c'a hwt'ae' hwdighitsigi.
/Food was really cheap.
12:38
C'aan łaneltsigi dollar-and-half nighile'.
/A whole sack of flour was a dollar-fifty.

Begin łaltsigi ce'i duugh nelbedzi yi łu' dollar quarter, dollar-and-half ghile'.
/A whole big slab of bacon this wide was a dollar twenty-five, (or a) dollar fifty.

T'ae' dighitsigi.
/Things were really cheap.

Yu' yu' n'sdelyaayi seł den, ts'iits'i tl'aseł udi'aan gaani,
/Clothing, clothes that we wore, the shoes, these called (ts'iits'i) denim pants,

yii su six bits 'eł ghile'.
/they cost six bits ($.75).

Dghaec tah łu yii c'a six bits, dollar hwk'e.
/Coats, too, were six bits or a dollar.

Some four bits ta ghile'.
/Some were four bits ($.50).

T'ae' dighitsigi.
/Everything was cheap.
13:22

(3)

Starting point: Chistochina
Ending point: Delta area
One-way distance totals about 160 miles.

Nggat Katl'abese' gha yet yii cu 'sneyeł.
/We would camp upland by 'up back riverbank' [5-24, hill on west bank of Chistochina R].

Yihwts'en xona 'unggat well . . . Nataghilen Na' yi cu 'sneyeł yet.
/From there upland to 'creek that water flows down' [5-29, East Fork of Chistochina R] there, too, we would also camp.

Yet cu łuk'ece'e cu una' talax.
/King salmon spawn it the stream there too.

Łuk'ece'e ts'eldiił ye c'a.
/We would eat king salmon there.
Noise
Nt'eyi?
/What? (to Katie)

Katie John:
Yae' ts'en gha snakaey 'iinn łnii.
/On that side [were] the children . . . he says.

Adam Sanford resumes:
Yeah. Yihwts'en xona 'unggat tez'aan.
/Yes. Up from there to upland [is] a fishtrap.

Yi nangge Nataghilen Na' ngga Tez'aani K'ae.
/Upland from 'creek that water flows down' upland is 'fishtrap hole' [5-30, fish camp on lower East Fork]
14:05
'Ahaan.
/Yes.

Tez'aani K'ae, yeah.
/'Fishtrap hole', yeah.

Duu yihwts'en Tsiis Tl'edze' Na' Ngge' 'stedeł dze' 'unggat Łedadlende.
/From here as we go to 'blue ochre river uplands' [5-23] upland to 'where streams join' [5-32, confluence of Middle Fork and Chistochina rivers].

(Someone asked)
"You want to ride horse?"[4]

KJ: Kedahwdełnesi.
/I hear that.

Du' yihwts'en nggat Tsiis Na', Tsiis Na' Łeke'udghidlende yi tey 'sdalts'iix.
/Then from there we stayed a long time at 'ochre river', 'ochre river' [5-32, Chisna R] 'where the streams join' [5-34, Powell Gulch].
14:33
Yii c'a Tsiis Na' yet hwts'en tey utl'aa hwnelt'e'i sut'e'i Tsiis Tl'edze Na'.
/That 'ochre river' [5-32, Chisna R] from there there are numerous headwater streams of the 'blue ochre river' [4-33, the Chistochina R].

Yihwts'en nggat Tsiis Na' kanaa 'utsii
/From there above 'ochre river' [5-33, Chisna R], across and down in the lowland

T'aghes Nuu Na' yii kugha ke'udghełen.
/'Cottonwood Island Creek' [5-34, West Fork of Chistochina R] flows past by there.

---

[4] Adam Sanford does not explain this, but this seems to be a reference to an encounter with men on horses going toward Slate Creek.

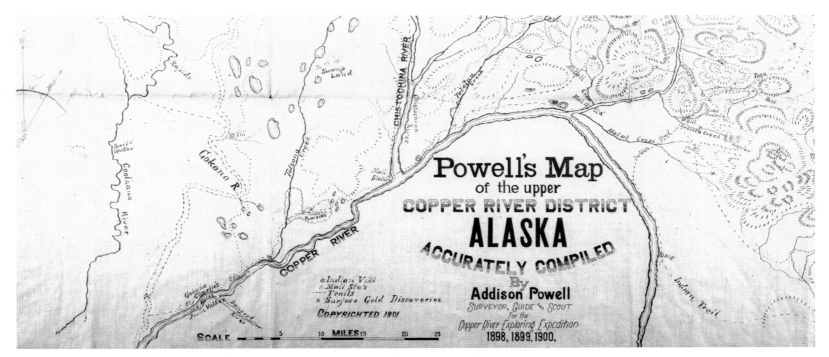

Figure 5-6. Southern portion of Addison Powell's 1901 commercial map of Eastern Alaska. This map is extremely rare; an original copy is on display at the Valdez Museum. This an important record of trails, place names, and Native settlements. Note fine details such as "Gakona Charlie's," trails between Gakona River [1-55] and Tulsona Creek [1-57] and "Tetelna" [5-54 **Di'idaedl Na',** Indian River], and trails between Suslota and the Batzulnetas areas. See also Figure 5-7, from the northern portion of this map.

Courtesy of the Valdez Museum and Historical Archive.

Yii c'a uluu' c'ilaen.
/It too has a glacier.

Yihwts'en Tsiis Na' Ngge' ye łu uluu' kol.
/In uplands of 'ochre river uplands' [5-35, Chisna R uplands] there is no glacier.

Tsiis Na'.
/'Ochre river' [5-32, Chisna R].

Big creek, you know, big river.
15:01
Du' yihwts'en xona Tsiis Tl'edze' Na' xona 'unggat.
/From there of 'blue ochre river' [4-33], to the upland.

Slate Creek, nts'e c'a k'alii nts'e koht'aene k'e k'alii kudziile?
/How come Slate Creek [5-36] is not named in Ahtna?

JK: Fred (John) could not remember either, Slate Creek. Gotta have a name.[5]

Yet hwts'en su xona 'utggu naan' 'utggu naan' kets'enaes dze'
/From there, then we go up above and across up above and across and

Ggax Ku .. Ggax Kutl'aa
/to 'rabbit *f.s.*' 'rabbit area headwaters' [5-37, upper Gakona R]

Ggax Kuluu' gha yet yii nits'enaes.
/and we stop at Rabbit area glacier' [5-38, Gakona Glacier].
15:37
Yihwts'en k'alii sut'e'e 'eł 'sdelts'iix xu.
/Beyond there we did not stay very comfortably.

Tsaani yaen' c'elaex, you know. Yii yaen' xu natel'as.
/There are just brown bears there. Only they roam around there.

Yii 'eł k'ali' sut'e'e uts'e' ts'ehwdeł'aha.
/We did not bother with them much.

JK: How about those mountains there, Adam?
Yea. Big glacier there. That mountain? Well, that mountain no call, I don't think, Indian no call, I guess.

---

[5] Slate Creek is one of the few larger sidestreams of the Chistochina River drainage with no remembered Ahtna name.

16:09
Yihwts'en ts'e' xona 'unggu naane xugha 'stenaes ts'en' above about Tanen'dalyaade.
/From there, then we would go across to there above is 'land between the waters' [5-39, the Alaska Range above the head of the Gakona R].

Xutah łuu xu'e'stediił.
/We would to amid there to the glacier area.

Yihwts'en about da'endze' diłeni yii tl'aa ts'enaes, ye du'.
/From there we came to the head of a stream flowing over to the other side (of Isabel Pass).

Tsiis Na' keniide.
/They call it 'ochre river' [5-40, McCallum Ck or College Ck into Delta R].

Deghilaay nełnaes. K'ał'aa deghilaay nełnaes.
/The mountain is tall. The mountain is truly tall.

Yii yii t'aax tsene ts'idiniłeni su, gha yet Tsiis Na'.
/It flows out from beneath them to the lowlands, there 'ochre river' [5-40].
16:46
JK: 'Uze' koley?
/It has no name?

Yea, 'uze' k'a c'ilaen xu xuk'a tey łdanahwdidlaeghe. Xosk'e'e 'ełaen tah łu.
/Yes, it may have a name but a very long time has passed. I am failing, it seems.

Gha yet Tsiis Na', Tsiis Deghilaaye' su tey deghilaay tey nilaen.
/By there, 'ochre creek', is 'ochre mountain' [5-40, 5-41, Cony Mt and group above Gulkana Glacier] is really quite a mountain.

Oh, high hill, you know.
17:10
T'ae' ts'abaeli kede'aayi k'a su k'ent'ae deghilaay.
/That mountain is just like a standing spruce tree.

Yet su ba'aadze' dae' du' Ketsiigi Na' ba'aaz dae' 'udi'aa xu.
/From the other side (Tanana River side) that way 'yellow river' [5-42, possibly Canwell Ck/Glacier] extends from there other side that way.

'Udaa'a nixuhnayeł una' 'eł tuu kole.
/Downstream they camp at a creek that has no water.
17:25
Naak yaen' ut'aghi'aa xunt'ae.
/Only rock bars extend through there.

Cu dangget Slate Creek Na', 'ungga Tsiis Na', yii yet Ketsiigi Na', ye c'a 'ałk'e gha kudzii you know.
/Note that up there Slate Creek [5-36, upland at 'ochre creek' [5-32, Chisna R and/or 5-40, McCallum Creek]]; 'yellow creek' [5-42, Canwell Ck], they are similarly named.[6]

Yet su xona yi na' tsene' ts'its'enaes dze'.
/There then at that stream [on Delta R] we came out toward the downland.

Ba'ene' Dzeł Ghatgge 'Ene'.
/Over at 'beyond between the mountains' [5-43, lower Delta R area].[7]
17:47
Ba'aa Łałuu'itsaak den Łałuu'itsaak den ye nits'enaes.
/Out at 'where glacier blocks an area' [5-44, Black Rapids Glacier], there we would stop.

Yihwts'en Una' Ketsitne'aay Na' yet.
/From there at 'its creek has a head against it' [5-46, Miller Ck] there.

Yedu' dae' ts'abaeli k'e su k'ent'ae 'utgga yii yet Una' Ketsitne'aay Na'.
/It is (steep), just like a spruce, above there, 'its creek has a head against it'.

Ts'abaeli nindez'aay k'e sunt'ae, nanuux 'utgge' xu tkut'ae.
/They (mountains) are standing like spruce [possibly Rainbow Ridge area] in the upstream area and above there.

K'ełt'aeni daagha.
/They are comparable to K'ełt'aeni [4-39, Mt Sanford or Mt Drum].
18:17
Yii t'aaxdze' nae' nits'enaes.
/We stopped out from beneath there to the upstream.

Xuxu deghilaay!
/Oh, those mountains!

Deghilaay tats'enaes ba'ene', Dzeł Ghatgge 'Ene'.
/We went among the mountains on the other side to 'beyond between the mountains' [5-43, lower Delta R area].

Ba'aa yihts'en ba'aat Kuułtaan Na' yii c'a duughe k'e hwt'ae' deghilaay nanilaa.
/Out from there from there out at 'area extends inside creek' [5-46, Bear Ck] the mountains extend across there.

---

[6] Adam comments on the duplication of and similarities to 'ochre'. See also Figure 2-2.

[7] The name **Dzeł Ghatgge,** 'between the mountains', is used for the Isabel Pass area.

Figure 5-7. Northern portion of Addison Powell's 1901 map.

The trail Adam Sanford describes in Part 3 is traced up the west bank of the Chistochina River, across Gakona River below the glacier, though a pass to Phelan Creek, and on to the upper Delta River.

Geoff Bleakley, historian at Wrangell-St. Elias National Park, in a summary of the transportation history of the Copper River and Wrangell Mountains wrote, "Alaska Natives established the area's first transportation networks. . . . These routes usually followed natural corridors such as river valleys and traversed the more obvious mountain passes" (Bleakley 1998:3).

By the early 1900s documentation on trails in the Copper River was very good and extensive. Some of the best sources are by guide Addison M. Powell (1901, 1910) and topographical engineer Oscar Rohn (1900a, 1900b, 1900c).

In the summer of 1898 Rohn traveled on foot and with horses circumambulating the Wrangell Mountains (Rohn 1900c). He also (1900b) describes 29 trails and passes along the Copper River and on the Tanana River side of Mount Wrangell. Rohn's phrasing on most trails is specific enough to indicate where there were visible existing Native trails. For example, Rohn (1900b:781) states: "An old Indian trail leads, in a general way, along the northern side of Tonsena River, reaching Copper about 8 miles above the mouth of the Tonsena."

By my count only four of the 29 trails listed in Rohn 1900b are referred to as developed or improved trails, and these are likely on the general course of pre-existing Ahtna or Upper Tanana trails.

See also Figure 6-3, the detailed 1903 Mendenhall and Schrader map of Chitina River.

Ba'aa Kuułtaan Na' yii cu nits'enaes. Yii cu 'unggu denyii tah.
/We stopped also out by 'area extends inside creek'.[8] Upland of there is a canyon.

Denyii tah yaen' ts'i'sdalts'iix.
/We just stayed in a canyon (of a sidestream).
18:40
Łts'ii, łts'ii hwditnes dze' da'andze' łts'ii łts'ii tah su.
/The wind, the wind continues (blowing) from beyond that side, it was a really windy.

Saen ta c'a xona koht'aene hwdlii taliił, you know.
/In the summer people can freeze, you know.

Xona ts'utsae xona ba'eni Dzeł Ghatgge 'Ene' kughile'.
/Well, long ago on the other side was 'beyond between the mountains' [5-43, the lower Delta R area].

Yihwts'en xona Ba'aadi Nene' K'e.
/From there (it went) is 'on the outside land' [5-44, the Tanana R Valley (general term)].

Dae' Łuu Tahwdzaeye' kiiłnii.
/That way they call it 'heart among the glaciers' [5-45, Donnelly Dome].

Yii ts'e' tsene tah. Yii detat'ax tsene tah 'sneyeł.
/From there is the lowlands. There in places we camped in the lowlands (of the Tanana River).
19:04
Yihwts'en xona 'utsiit, nts'e kol t'aen?
/From there, then toward the low land, how can it (the place name) be absent?'

Xona Nizaayh Caegge.[9]
/Then is 'gravel mouth' [5-46 the mouth of Jarvis Ck].

Nizaayh Caegge ye nits'enaes.
/We stopped there at 'gravel mouth'.

Yits'en tle xona, xona Ba'aaxe Tuu', Ba'aaxe Tuu' baaghe dii ts'its'edeł.
/From there a little bit, then 'outside water' [5-47, the Tanana R] the shore of 'outside water', there we come out.

---

[8] Inferred location is Bear Creek, based upon the meaning of **Kuułtaan** 'area extends inside', as this stream seems to offer the most wind protection in this district.

[9] Adam seems to be using a voiceless final front velar yh here, not typical of Ahtna but found in the Middle Tanana language.

Xu' łu' xona ba'aaxe hwt'aene tah ghats'edeł.
/There then we came among the Tanana River peoples.

Xoxoxo koht'aene una' c'ghile' Ba'aaxe Tuu'.
/Oh, there were people on the streams, on 'outside water' [5-47, Tanana R].

Du' łu hwna koht'aene 'eł 'sdalts'iix.
/Meanwhile we stayed among the people.[10]

Xayde ba'aadze c'uka na'stadeł.
/In the winter we would come back hunting from the other side.

Niłdentah t'ae' nse'dze' xuk'a na'stadeł.
/Sometimes we would return later toward spring.

Duugh łu' tsa' dzaghe' diłcax xu duugh nanats'adeł.
/There where that 'beaver ear' (a plant, wintergreen, Pyrola sp.) was only so big when we were returning.

(We would) walk all the way, ggaał kae
/with speed (power walking).[11]

Cu gaani k'e 'sneyiił xona ba'aadze.
/Again here then (on Copper River side) we camped then (coming) from the other side.

Teye kudesaat de xu su tkut'aede Dzeł Ghatgge 'Ene'.
/It is really a long distance, to there to 'beyond between the mountains' [5-43, lower Delta R area].

K'adii łu car su dae' kudułdiiye de.
/Now it seems with a car it has become a short distance.
20:11

(4)

This interlude section was omitted from the 1986 version.
Du' xona xa' open na 'ilaen.
/Well, then open it (tape recorder).

---

[10] Adam did not specify where, but he likley was at Goodpaster Village or perhaps as far downstream as Salcha Village.

[11] Cf. Kari 1986:174. The term **ggaał** is an Ahtna homonym that means with 'alacrity, speed' or 'snare.' My translation (with 'snares') in 1986 was wrong, as obviously they were not walking back and setting snares.

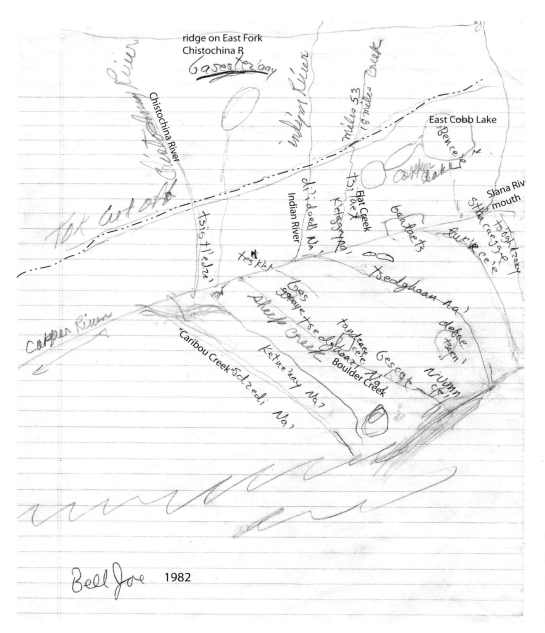

Figure 5-8. 1982 sketch map by Bell Joe of the Copper River in the Chistochina area.

While drawing this map, Bell mentioned about 20 place names, nine of which have been mentioned in Chapters 1, 4, and 5. Bell Joe's detail for the Boulder Creek area was a nice complement to the names provided by Adam Sanford.

K'adii dzaene k'adii dzaene koht'aene k'ehnaesen c'a sa na'idyaa.
/Today, today the one who speak Ahtna returned to me.

Dae' du' u'eł szaa tkonii, u'eł szaa tkonii xona.
/Thus with him I am talkative (lit. I have a mouth), I am so talkative with him.

Koht'aene k'ehnaesen gaa tseh k'a u'eł 'estnes gaa su na'idyaan.
/The one who speaks Ahtna here I know him previously and he has returned here.

Hwt'ae' good heaven nada'sdełtsiin.
/We have made a good heaven verbally.

Ugheli uts'en daghatlaeyi nełnii
/Good things from it (this recording) should occur, he tells us.

K'adii yen si c'a k'adii u'eł kensyaesi.
/Now it is him with whom I am speaking.

Tsin'aen k'adii nanghal'aen.
/Thanks to see you again now.

That's enough.
JK: Tsin'aen.
20:47

(5)

Starting point: mouth of Sanford River (1) down to Gulkana, toward Gulkana River uplands
Ending point: Sanford River to upper Nabesna River
Each route is about 80 miles.

'Udaat 'udaa Sngadaxi daagge' yet yedu' yedu' den nisiltaen den.
/Downriver, upon 'sliding' [5-48, hill on Copper R above mouth of Sanford R], right there I was born.

Yet ts'insyaade sii st'aet.
/That is where I came from when I came of age.

Yihwts'en Dzaan Yighilende yet su xak'a k'ensdzet.
/From there to 'where it flows into murky water' [5-49, site near mouth of Gulkana R], right there I grew up.
21:06
Enaa! Duugh!
/Oh my! Around there!

Snaan łu duugh unaan k'et nadaat Ts'itaeł Caegge hdaghalts'e' den.
/Then my mother and her mother stayed at 'mouth of river that flows straight' [5-1 north of Sanford R].

Xona snaan łu, she come. Denaan 'iin ts'e' natesgaa.[12]
/Then my mother, she come. She went back to her mother's people.

Xugha na'idyaa naene c'a naene c'a dents'ełtsigi. T'ae' dghałtsigi.
/When she returned to them, we were small. I was really small.

Well, C'ulc'ena' Ngge' 'utggu łu'snidaetl you know, C'ulc'ena' Ngge'.
/Well, we traveled the 'tearing river uplands' [5-50 Gulkana R uplands], the 'tearing river uplands'.
21:36
Yet hwts'en 'unggat Taltsogh Caegge yet cu neghak'ae kughile', a little while.
/From there, upland at 'brown water mouth' [1-58, Tulsona Ck] our home also was there for a little while.

Yi cu łu łuk'ae gha 'sdalts'iix.
/We used to also stay there for salmon.

Tsabaey, tsabaey gha 'esdalts'iix.
/Trout, we used to stay for trout.

Yet hwts'en ngge' saene tene na' Saen Tene Ngge' tanats'edeł 'unggu.
/From there we used to go upland on the summer trail to 'the summer trail uplands' [5-51, upper middle Gulkana R and area toward Tangle Lakes] upland there.

Sii du' t'ae' dghałtsigi.
/I was very small.
22:00
Cu xona sta' 'iine nen' ta tene kughił'aen.
/My father's people had a trail out into the country there.

'Unggu xona deghilaay tanats'edeł.
/Then we used to go up into the mountains.

Tseles ugha ghat'aen'i.
/That is a place for getting ground squirrels.

Tseles hghiłkaan' dze'
/They liked the taste of ground squirrels and

---

[12] **Gaa** and not **dyaa,** the lower dialect, perhaps in imitation of his mother's speech.

hwt'aene koht'aene yii su nen'ta kiigha dghat'aen'i, tseles.
/all the people went into the country for them, ground squirrels.
22:16
Uyiit gha tketl'uux. Tseles nkełtaes du' yeden xu yedi'i ggaay ye xu'.
/In there they set snares for them. They harvested ground squirrels, whatever other little things are there.

Udzih c'a cu hzełghaes, yii kae c'a xu nen' tah.
/They killed caribou, too, (living) on that out in the country.

Hwt'ae' 'ele' sut'e' c'a ugheli hghilaele nen' kughile' ts'utsae.
/They did not have a very good time, the way the country was long ago.

Hwt'e' tseles c'ghile'. Du' udzih c'ghile'.
/There were just ground squirrels. There were caribou.

Deniigi yet k'a kughistle'e. Nen' ghiłcaaxdze' kughistle'e.
/There weren't any moose. In the entire country they were absent.

One or two k'a ghidaax.
/Only one or two were staying.

Ts'ełk'ey c'a nen' k'e ghidaax, you know.
/One might be staying in the country.
22:51
Yii łu nakiłtaesi 'eł hzghiłghaes, yii kae.
/If they found it, they would kill it, with that.

K'adii su deniigi c'ezdlaen.
/Recently moose have appeared.

K'adii da'atnaey 'ełaen tah xona, denae . . . [false start] deniigi nen' k'et c'ezdlaen.
/Now, when I am an old man, then the moose have appeared in the country.

Yii cu u'eł 'estnes.
/I know that.

 Du' xona yihwts'en C'ulc'ena'
/Now from there, up from 'tearing river' [1-51, Gulkana R]

'ungga k'adii c'etsiy tnaey hdelts'ii de yet su k'adii kayax kuzdlaen.
/upland where now the white people stay [Gakona area], now it has become a village there.

Figure 5-9. An officially unnamed mountain, elevation 6,580 feet, east of Mount Sanford and north of Copper Glacier.

The Ahtna name, **Siz'aani** or **Sez'aann** [4-38], is literally 'inside me' and is often translated as 'my heart'. Copper Glacier is called **Ts'itu' Luu'** 'major river glacier'.

Photo by University of Alaska Fairbanks, Geophysical Institute, courtesy of Carl Benson.

23:20
C'etsiy tnaey gha tseh su 'ut . . . 'utsiit Dzaan Yighilende yet kughak'ae kughile'.
/Before the white people in the low land at 'where it flows into murky water' [5-49, near mouth of the Gulkana R], that was their home.

Yet su nitsiił 'eł ghile'. Yet'aax hdalts'iix xayde.
/There were winter houses there. They stayed in there in the winter.

Du' yihwts'en naene xona Ts'itaeł Caegge kanats'esdaetl.
/From there, then we went back up to 'mouth of flows straight' [5-1].

Stsucde 'iin, stsucde 'iin gha nats'idaetl.
/We returned to my grandmother's people, my grandmother's people.
23:38
Gaa su Ts'itael Caegge kanaa 'uniit Sdaghene sdates 'ene' tighita'.
/Here at 'mouth of flows straight' [5-1] across from and upstream at a 'point river bend' [5-52, site at bend or point below Sdzedi Na'] there was a trail up over the bend.

Kanaa 'uniit xona hwnax ce'e kughile'.
/Across and upriver was a big house.

Ts'utsae hwnaxe' nitsiił ye ghi'a'.
/An old-style house, a winter house was there.

Yii nae' nats'adeł xayde.
/We went back upstream there in the winter.

Yi daagge' xayde 'sdalts'iix.
/We stayed on that bluff in the winter.

Du' yihwts'en duu daadze' na'stedeł dze' xona duugh.
/Then from there, coming from downriver, we came back around here [to Chistochina area].
24:04
Duu 'udaadze 'stedeł dze' 'uniit Sdzedi Caegge yet 'sneyeł.
/We come from the downstream and upriver here at 'sdzedi mouth' [5-53, mouth of "Caribou Creek"], there we would camp.

Yet Taltsogh Caegge cu nats'udeł.
/Or then we might come back again to 'brown water mouth' [1-57, Tulsona Ck mouth].

Cu yet ts'eneyeł Taltsogh Caegge, Sdzedi Caegge c'a.
/We camped there again at 'yellow water mouth' or at 'sdzedi mouth' as well.

Yihwts'en xona Bancdidaasi Tayene', dae' ts'en ta ba 'snighiyeł.
/From there at 'straight stretch that game cross' [1-61, on Copper R near mile 29 of Tok Cutoff] we camped by it on that side.

Yet yihwts'en nanaa 'uniit Ts'itu' K'et, Ts'itu' K'et dinats'adeł.
/From there across and upriver to 'on major river' [5-21, fish camp near Chistochina] we come back to 'on major river'.
24:31
That fish camp over there.

K'adii łuk'ae gha 'sdelts'iix.
/Now we stay there for salmon.

Yet su Ts'itu' K'et hwdghi'a'.
/That was called 'on major river'.

Duu yihwts'en naniit Di'idaedl Na' yet xona uc'aadze' xona uc'aadze' kiidiłeni.
/From there upriver to '(fish) go in river' [5-54, Indian R] then opposite from it, opposite from it a stream flows in.

Yea. Snał'ii nahwdalzes.
/Yes. That name has slipped my mind.

Kanaa uc'aadze' cu yet cu kayax kughile' kanaa uc'aadze'.
/Across and opposite it there also there was a village, across and opposite.

K'adii Ts'ensdzedi Na' cu.
/Across from there also is 'we turn back unsuccessfully creek' [5-55, creek into Copper R opposite mouth of Indian R].
25:07
Du' yihwts'en xona ts'inats'atnaes 'uniit.
/From there then we would start to move out again to far upstream.

Nuu T'aax Dezdlaade yet 'snighiveł.
/We camped there at 'objects beneath the islands' [5-56, on Copper R near Cobb Lakes].

Du' yihwts'en ts'inats'atnaesi Baa Łaedzi Cii 'snighiveł.
/Then out from there we would start to move out again and we camped at 'grey sand point' [1-67, point near Slana].

Yihwts'en xona 'unuu Uk'e Nic'ahwdetsedzi K'et 'snighiveł.
/From there, then upriver we camped at 'on it dry wood goes out from shore' [5-57, point on Copper R on south bank above Slana].

Yet ts'en di Nataełde kanats'adeł.
/From there we came back up to 'roasted salmon place' [1-69, Batzulnetas].
25:36

This next 30 sec. was accidentally omitted from the 1986 version.
Oh oh tayen dighiłnaesi yet kut'aet
/Oh the straight stretch is very long.

'Uniit 'uniit Nataełde yet tayen ts'ini'aayi su 'udaat 'udaat Stl'aa Caegge k'a keghi'a'.
/Upstream, upstream at 'roasted salmon place' [1-69] a straight stretch extends out distantly downstream, downstream to the mouth of 'rear river' [1-68, Slana R mouth].

'Udaa yet su C'alts'iis Bese' dae' du' hwdghi'a' den.
/Downstream there it is named 'rough place river bank' [5-58, bluff at Ahtell Ck mouth].

Figure 5-10. North side of Mountain Wrangell from the Slana area. The main name for Mount Wrangell and the Wrangell Mountains is **K'ełt'aeni** [1-25]. The steaming volcano at Wrangell's summit is called **Uk'ełedi** 'the one with smoke upon it.' The belief of the traditional Ahtna, Upper Tanana, and Tanacross people is that the souls of the dead can be viewed in the volcano's smoke.

Photo by Suzanne McCarthy.

Yak'a kaghi'aa tayen ts'ełk'ey tayen k'a ghile'.
/That extends up there, that straight stretch, that is just one straight stretch,
25:52
Uk'e kudesaat gha tayen.
/It is a long distance upon it, that straight stretch.

'Unaa 'uniit ts'ihwniidze xa'a yełu Uk'e Nic'ahwdetsedi.
/Distantly across, distantly upstream right in the middle is 'on it dry wood goes out from shore' [5-57].

Sdaa su 'adetnii.
/That is called a peninsula.

Sdaa uk'et tsets nuu k'et c'ghile' Uk'e Nic'ehwdetsedzi.
/At the peninsula dry wood was on the island there, 'on it dry wood goes out from shore' [5-57].

Yet 'snighiyeł ts'inse Nataełde kants'adeł.
/There we camped and straight ahead we come back to 'roasted salmon place' [1-68, Batzulnetas].

26:10 resumes, and included in 1986
Yet hwts'en dangge saen tah kanats'adeł niłdentah.
/Upland from there is a trail to uplands in summer in some places.

Yihwts'en xona danggeh xona nen' nen' tah teni.
/From there then in the uplands then is the trail out to the country.

Sez'ae 'iin yet hdaghalts'e' ne 'eł nen' ta'stadeł.
/My uncle's people (Sanford Charley) stayed there and we would go out into the country.

'Ungga Ts'abaeli K'edi k'et yet niłdentah 'snighiyeł.
/Sometimes we camped upland on 'the one with spruce on it' [4-22, hill east of Batzulnetas].

Niłden 'unggat K'eseh, K'eseh 'snighiyeł.
/Sometimes far upland at the lake outlet [5-59, outlet of Tanada L], we camped at 'the outlet', 'the outlet'.

Yet 'ungga Men Diłeni Caegge su yet c'a yi na' 'snighiyeł.
/We also camped upland at that creek there, 'mouth of that which flows into the lake' [4-24, creek into southeast shore of Tanada L].

Yihwts'en xona, xona deghilaay yits'enaes.
/From there then, then we moved into the mountains.

Xona niłk'ae gha deghilaay t'ae' deghilaay yaen' tah.
/Then there are mountains on both sides (of Tanada L), only mountains.

'Unggat Dzahnii Bene' nidinił'a' de.
/Upland of 'rarely said lake' [5-60, Copper L] where current flows to a place.
26:53
Duu yits'en xona dangget Nitsic'ełggodi TI'aa hwts'e'.
/There from there then to the upland place 'rock is chipped headwaters' [4-28, upper Jacksina Ck].

Xona 'stxinaes de xu debae ggaay kae.
/We went there (living) on small sheep.

Xu Debae ggaay k'alii sut'e' c'a debae c'ghilaele xu łu.
/The small sheep weren't very good. Not much sheep.

Not much (game) in there, you know.

Xona ye k'ae su xona cu Nitsic'ełggodi Tl'aa ts'eghinaes.
/Then there we came out from 'rock is chipped headwaters' [4-28].

Yihwts'en k'a kana'snighinaes.
/From there then we would turn back.

Gha yet 'unggat łuu gha, łuu gha hwts'e' kudełdiyede gha kanat . . . kana'snighideł.
/From there to the glacier [5-63, Nabesna Glacier], we turned back just a short distance from the glacier.
27:29
Katsuughe yaen' łi cu Tsae T'aax 'unaa yi c'a 'ungga łuu gha yet c'a ts'ekaltiin'.
/In the lowland area at 'beneath the rock' [4-31, Wait Creek Pass], across there they had a trail
out upland to the glacier.

Nabaes Na'.
/'Types of stone river' [5-64, Nabesna R].

Yet ts'en su xona 'unggase nen' nen' tiisde gha ts'ents'edeł.
/From there then we come back from the uplands as the ground freezes.

Xu Debae gha saen hdalts'iix xu.
/They stayed there during the summer for sheep.

Dae' tkat'aen'i kae su. Nen' k'e 'sdaghalts'e'de.
/That is what they lived on. We stayed out in the country.

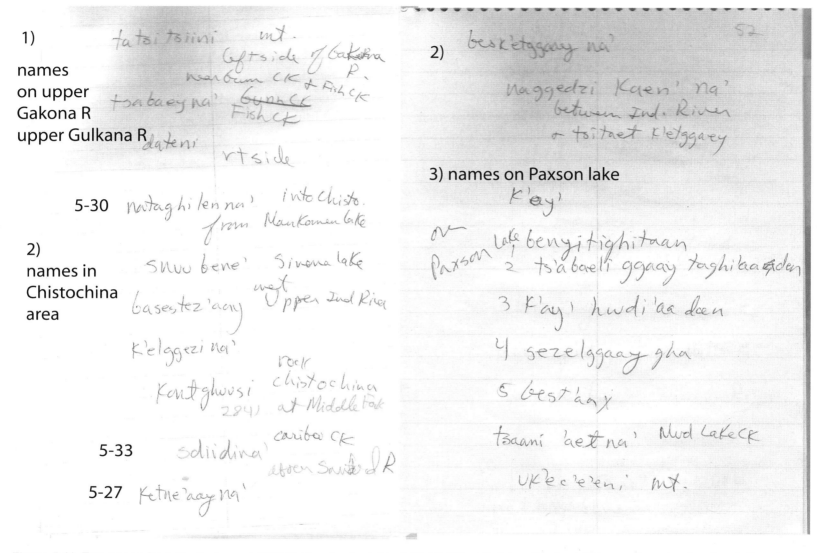

Figure 5-11. Two pages from Kari's August 1978 field notes (notebook #6:50–53) with Adam Sanford. This was the only other time before June 13, 1981, that we worked extensively on place names. In the 1978 session Adam mentioned 39 place names, nine of which are also in his 1981 narrative. Several names could never be reconfirmed by other Ahtna speakers. For one such name at the top left, **Tatsitsiini,** Adam noted the location fairly precisely, and I now think that it is the ridge north of Gunn Lake. Every listing of names by a great expert is valuable.

T'ae' ne'ahwdighitiy'.
/It was very hard for us.
28:00

(6)

K'adii su 'sc'ezyen ta 'eł xu dyaak.
/Now it has happened that we eat well.

Ts'utsaede t'ae' ne'ahwdighitiy'.
/In the past it was really hard for us.

Ghayuu nts'e tkot'aen da c'etsen' ggaay 'eł łu yuugh niłts'e' koht'aene gha c'ghila' kole.
/How they did with just a little meat, and there was not much to share among the people.

K'alii koht'aene gha c'ilaale.
/The people did not have much.

Just kiiyaani kae su nen' k'e hdaghalts'e'.
/They just lived on the land with what they could eat.

Nen' k'e łuhninidaek.
/They went around the country in poverty.

Du' xona xona k'adii c'etsiy tnaey nen' kezdaedl 'eł xona
/Well, now, now that the white people have come, and so
28:32
xona c'aan 'eł dghitaey' xona.
/so there is enough food then.

Xona lzaasi 'eł c'ezdlaen xu.
/Then money appeared.

Xooxoo k'adii lzaasi k'adii c'a k'adii c'a da ukol koht'aene.
/But oh, now the (Ahtna) people have no money.

Koht'aene ta'itsaak xudyaak.
/It (money) came among the people.

C'etsiy tnaey xona k'aatle c'etsiy tnaey k'a kezdlaen.
/White people, so then they (the Ahtna) have almost become white people.

Figure 5-12. Swans (kaggos) **taking flight at** Hwdagguus Bene' **'celery mouth lake' (Miers Lake).** Photo by Suzanne McCarthy.

C'etsiy tnaey kezdlaen.
/They have become white people.
28:29
K'adii gaat sk'ehnaesen sa na'idyaa dze' szaa tkonii.
/Now here the one who talks like me came back to me and I am talkative.

Tsin'aen nanghal'aen.
/Thanks seeing you again.

Uk'e k'a kensyaes. Xona.
/I speak to him. That's all.
ends 29:08

6

Summary of Ahtna Place Names and Riverine Directionals

Ahtna Riverine Directionals

As the Ahtna experts present these walking tours, several complex linguistic tools being are employed. This *shared geographic knowledge* has many regular and formal features. Four features of Ahtna geographic names contribute to the learning and memorization of the geography: *name content, name structure, name distribution* and *name networks* (Kari 2008:15–31, to appear). In addition, Ahtna and other Northern Athabascan languages are oriented towarde the major rivers, and they employ an elaborate *riverine directional system*. The Athabascan riverine directional system is the organizational intersection between the geography, the lexicon, and the grammar.

Riverine directional terms are comprehensive and pervasive in Ahtna. They are used in indoor and outdoor settings, and in several vocabulary and grammatical categories (place names, parts of houses, anatomical terms, verb prefixes). The riverine directionals in Athabascan languages are what Levinson (2003:90) terms an "intermediate absolute landmark" frame of reference. The major rivers can have totally different geographic axes such as the Copper River (which flows in an arc north to south) versus the Tanana River (which flows east to west). The directionals are accompanied by gestures and body movements to or from the major river (usually the Copper River), a great topic for further study.[1]

In Ahtna and other Alaska Athabascan languages the directionals are a special word category with prefix + root + suffix structure. This structure resembles the verb complex in miniature, and it occurs typically in more than 60 derived forms. Table 6-1 displays the morphology of the Ahtna riverine directionals with the prefixes and suffixes that combine with the set of nine roots.

---

[1] Northern Athabascans do not use fixed cardinal directions (north-south-east-west) for orientation. Terms for left and right are not used at all for spatial orientation or in geographic terminology.

meanings	plain ADV	5 prefixes	bound roots	5 suffixes: -Ø, -xu, -t, -dze', -ts'en
across	naane	ts'i- 'straight'	-naane	-naane, -naaxe, -naat,-naadze'
downstream	daa'	'u- 'far'	-daa'a	-daa'a, -daaxe, -daat, -daadze'
upstream	nae'	da- 'close'	-n'e	-n'e, -nuuxe, -niit, niidze'
lowlands, toward water	tsen	ka- 'adjacent, next'	-tsene	-tsen, -tsuughe, -tsiit, -tsiidze'
uplands, upland area, from water	ngge'	na- 'intermediate'	-ngge'	-ngge', -nggu, nggat, -nggadze'
up	tgge'	P+gha 'in relation to'	-tgge'	-tgge', -tggu, -tggat, -tggadze'
down	yax		-ygge'	-ygge', -yggu, -yggat, -yggadze
outside, beyond, other side	'ane'		-'ane'	-'ane', -'aaxe, -'aat, -'aadze'
forward, ahead, front, to fire	nse'		-nse'	-nse, nse', -nsghu, -nset, -nsedze'
totals	9 + 9 x 6 x 5 = 69 plain or derived directional forms

Table 6-1. Ahtna riverine directional structure.

With the plain adverbials (without prefix or suffix) and the combinations of prefixes, roots and suffixes there are sixty-nine possible directional terms. Expert speakers can tap into all of these fine distinctions. Also, there are morphophonemic alternations typical of the verb stem variants, such as **tsen + t -> tsiit, 'an + xu -> 'aaxe,** and **n'e + t -> niit.** Short of having a morphemic gloss for every word, it is difficult to translate the precise meanings of derivative directionals:

natsiit   'close + downland + at specific place'
'unuuxe   'far + upstream + in general area'

Suffixes can reverse the path of directionals: **kanggadze'** 'from the next place upland' contrasts with **katsene** 'toward the next place downland.' In this book we try to present repeated uses of the directionals, but we have not attempted to give the precise meanings of each derived directional term (although the Ahtna experts certainly distinguish more than sixty precise meaning with these words).

The Ahtna directionals provide a remarkably comprehensive frame of reference. In Figure 6-1 and Table 6-2 we summarize the semantic oppositions and applications of the directionals.

In Figure 6-1 and Table 6-2 it is interesting to see how the nine roots are in *orthogonal oppositions;* that is, they are mutually perpendicular. Seven of the nine roots have the same orthogonal oppositions both indoors or outdoors (in the local landscape) as in settings 1, 2, and 3 in Figure 6-1: across :: across, upstream :: downstream, upland :: downland, and up :: down (vertically). The entrance to the traditional Ahtna house faced

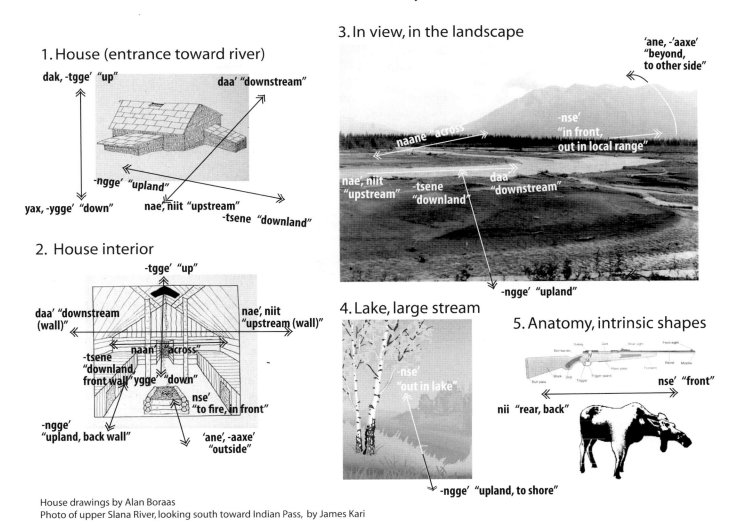

1. House (entrance toward river)

dak, -tgge' "up"

daa' "downstream"

-ngge' "upland"

yax, -ygge' "down"

nae, niit "upstream"

-tsene "downland"

2. House interior

-tgge' "up"

daa' "downstream (wall)"

nae', niit "upstream (wall)"

-tsene "downland, front wall"

naan' "across"

ygge' "down"

nse' "to fire, in front"

-ngge' "upland, back wall"

'ane', -aaxe' "outside"

3. In view, in the landscape

'ane, -'aaxe' "beyond, to other side"

naane "across"

-nse' "in front, out in local range"

nae', niit "upstream"

-tsene "downland"

daa' "downstream"

-ngge' "upland"

4. Lake, large stream

-nse' "out in lake"

-ngge' "upland, to shore"

5. Anatomy, intrinsic shapes

nse' "front"

nii "rear, back"

House drawings by Alan Boraas
Photo of upper Slana River, looking south toward Indian Pass, by James Kari

Figure 6-1. Ahtna directionals, composite view.

gloss	root	gloss	root	applications
1) across	naane ⇔	across	naane	outdoors or indoors
2) downstream	daa'a ⇔	upstream	n'e/nii	streams, sidewalls of house (entrance facing stream)
3) downland	tsene ⇔	upland	ngge'	to/from water body, front/back wall; stream drainage
4) up	tgge' ⇔	down	ygge'	outdoors or indoors
5) to front, to fire	nse' ⇔	outside, over	'ene	indoors: to fire/outdoors; open space: in open, to front/over, beyond; interregional (over mountains to the south :: over mountains to north)
6) out in water	nse' ⇔	toward land	ngge'	lakes, larger streams
7) front	nse' ⇔	back	nii	intrinsic shapes (boat, rifle) anatomy (human, quadraped)

Table 6-2. Seven semantic oppositions in nine directional roots.

the local stream, so the front wall is 'downland' and the back wall is 'upland'.

The root **nse'** is the most interesting and semantically complex directional root. In Figure 6-1 and Table 6-2, 5-7 **nse'** occurs in three pairs of oppositions. Inside and outside (settings 1, 2, and 3) **nse'** patterns opposite **'ane'**. The position of the hearth is the semantic thread for, **nse' :: 'ane'**. The hearth was in the center of the house, and persons face the fire in both indoor and outdoor settings and at three distinct scales: (a) *indoors*: to the fire versus outdoors; (b) *out in the local range*: into the open, to the front versus over, beyond (a hill, mountain); and (c) *long range* (over Chugach Mountains to the south :: over Alaska Range to north) to the fire, to the front (see Figure 6-2).

Rounding out the semantic oppositions for **nse'** in Figure 6-1 and Table 6-2, for intrinsic shapes (boats, any shaped object) and anatomy **nse'** is 'front' and **n'e, nii** is 'back', thus the dual meaning for **n'e, nii** is 'upstream; back.' In lakes and large streams the opposition is **nse' :: ngge'** 'out on water; to the shore, to the upland'.

Ahtna speakers have a convenient way of depicting directions when they refer to the neighboring regions of Alaska. We can visualize the long-range opposition of **nse' :: 'ane'** in Figure 6-2 on an oblique landform map of Alaska.

Some Uses of Place Names and Riverine Directionals in the Five Travel Narratives

The Ahtna and the Northern Athabascans are among the world's foremost pedestrian foragers, and they speak languages that are highly unusual and complex. The features of the travel narratives that the speakers use for navigation and orientation are topics that invite further research. Having a good corpus of mapped travel narratives is an essential step toward a formal study of the use of Ahtna and Athabascan geographic names

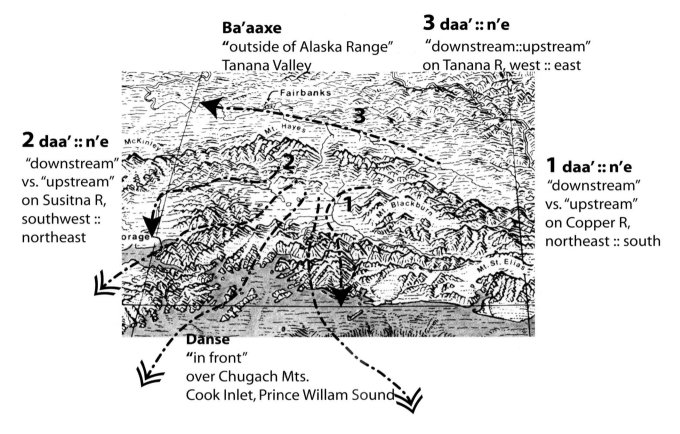

Ba'aaxe
"outside of Alaska Range"
Tanana Valley

3 daa' :: n'e
"downstream::upstream"
on Tanana R, west :: east

2 daa' :: n'e
"downstream"
vs. "upstream"
on Susitna R,
southwest ::
northeast

1 daa' :: n'e
"downstream"
vs. "upstream"
on Copper R,
northeast :: south

Danse
"in front"
over Chugach Mts.
Cook Inlet, Prince Willam Sound

Figure 6-2. Ahtna directionals at the regional or long-range scale. Depicted are upstream :: downstream for the different trajectories of the three major rivers in the region: the (1) Copper, (2) Susitna, and (3) Tanana. A regional place name, **Danse,** refers to the direction that arcs to the south over the Chugach Range. It appears that the southerly direction **Danse** is based upon 'in front, toward the hearth' in the sense that one is facing the "hearth" as the arc of the sun in the southerly sky. Opposite this is **Ba'aaxe** 'outside', which refers to an arc to the north over the Alaska Range to the Tanana Valley.

and directionals. In this section we summarize some of the uses of place names and directionals in the five narratives.

In these narratives the geographic names are fixed lexical items that occur in sequences whereas the directionals are pragmatically applied to orient and clarify directions and areas. Table 6-3 is a summary of place names and directionals for each chapter and segment. These are the first figures of this kind for a corpus of Athabascan travel narratives.

Tracking the Use of Place Names

Redundancy, repetition, and reconfirmation of geographic particularism are the hallmarks of Athabascan geographic knowledge. To better understand types of redudancy we have defined some terms, many of which can be tracked in the set of travel narratives in Table 6-3, Part A.

(a) *Distinct place names* (names for places that are distinct locations): total 313

(b) *Repeated place names* (the same speaker repeats a place name): total 698. Jim McKinley has a "lecture" style. As he mentions a place and explains why it is so named, he repeats a place name several times. In the most detailed walking tours by Jake Tansy and Adam Sanford, repeated place names usually reflect alternate routes and trail junctions.

(c) *Specific names vs. generated names,* as when generic terms—'stream mouth' 'headwaters' 'riverbank,' 'hill'—combine with the specific name to form generated names. The 313 distinct place names in five chapters were generated from 237 specific names. In other words, 23.5 percent of the names were generated. We refer to this as the *Athabascan generative geography capacity.* This is an important demonstration of how this capacity works in spontaneous speech that describes the Ahtna landscape.

(d) *Sets of names* with the same specific name. See Figure 4-4 for a set of eight place names with **tak'ae** in Mentasta Pass.

(e) *Duplicate place names* (different geographic features that have the same name). See Figure 2-2 for three examples.

(f) *Reiterated place names* (different speakers mention the same place names). In the five narratives, the repeated place name reference numbers occur 41 times. This overlap occurs in Chapters 1, 4, and 5. However, throughout the book we have used many other distinct and reiterated Ahtna place names in historic maps, sketch maps by Ahtna speakers, field notes, photographs, and essays. At the bottom row of Table 6-3 is a tally of another 292 distinct names (not otherwise in the narratives), of which 162 are reiterated names. These figures show the some of the ways that corroboration, reconfirmation, and redundancy occur when different Ahtna speakers have reported geographic names, perhaps in the 19th century on maps and historic records, or in the context of research.[2]

---

[2] Note that three other sources on Ahtna place names that have not been cited in this book—deLaguna 1970, West 1973, and Reckord 1983—contain perhaps 200 of the the most well known Ahtna place names, and almost all of these names have been reconfirmed by numerous Ahtna speakers.

chap (sec)	time	route miles	A. Place Names					B. Directionals									
			dis- tinct p.n.	spec./ gen.	total p.n.	reit. p.n.	Eng p.n.	total dir.	n'e upst.	daa' dnst.	tsen dnld.	ngge' upld.	tgge' up	ygge' dwn.	nse' front	'ane out	naan across
1(1)	27:34	170	72	64/72	229	0	2	141	61	13	0	10	4	3	6	3	39
1(2)	17:47	110	46	35/44	146	0	10	90	2	1	9	42	18	0	5	5	8
1	45:12	280	**118**	99/116	**375**		12	**231**									
2	6:40	80	**11**	7/10	**26**	0	9	**11**	4	0	1	3	1	0	0	1	1
3(1)	3:12	120	22	16/22	36	0	0	26	0	4	5	5	0	0	1	5	2
3(2)	5:12	80	18	16/18	43	0	4	32	5	6	4	5	2	0	2	9	4
3	8:24	200	**40**	32/40	**79**			**87**									
4(1)	4:40	25	21	11/21	34	9	1	23	0	0	7	10	0	2	0	3	3
4(2)	4:50	80	23	19/23	29	3	1	29	1	2	5	14	2	0	0	4	1
4(3)	2:30	35	12	12	17	7	2	17	4	5	4	0	1	0	0	2	1
4	12:00	140	**56**	42/56	**80**			**69**									
5(1)	9:24	110	25	14/25	37	3	0	67	10	2	3	30	10	0	0	3	9
5(2)	3:20	50	10	4/10	14	8	0	28	4	0	0	14	1	0	1	1	7
5(3)	6:20	160	23	16/23	46	3	1	50	1	1	6	16	4	0	0	19	3
5(5)	5:05	160	30	28/30	41	8	0	59	10	8	2	18	1	0	1	2	7
5	23:45	480	**88**	62/88	**138**			**204**									
totals	102: 11	**1180 mi.**	313	237/ 24.3%	698	41	40	**573**	102	42	38	167	44	5	17	57	85
p.n. in essays, maps, captions, field notes			296			164												
total distinct vs. reiterated place names			**609**			**205**												

Table 6-3. Summary of the use of place names and directionals in Chapters 1–5 of this book.

It is also interesting to see how speakers use local Alaska English versions of place names in relation to Ahtna place names, in the column *Eng. p.n.* If the speaker uses "Klutina" instead of **Tl'atina'**, that was tallied as an English name. The expert speakers can use the language purely; Chapters 3, 4, and 5 have only nine Alaska English place names. Tracking the use of local English place names reinforces the concept that the Ahtna geographic names form a truly autonomous conceptual system.

The Use of Directionals

Tracking the use of directionals in the narratives requires a consistent policy for distinguishing directionals that occur in place names, from those that are being applied pragmatically to clarify direction and space. Not counted as directionals in Table 6-3, part B are several types of place names that include directional terms such as:

Hwdaadi Na'	Dadina River	'downriver river'
Hwniidi Na'	Nadina river	'upriver river'
Hwtsuugh Naknelyaayi	mt 4716' S of Denali Highway	'lowland ridge-that-extends-across'
Henggu Naknelyaayi	mt above 4716'	'upland ridge-that-extends-across'
Tl'atina' Ngge'	Klutina River drainage	'rear water uplands'

In the five chapters we count 573 directionals (in 102 minutes of speech). All nine of the directional roots are attested, which verifies the comprehensive, multisetting, three-dimensional oppositions of the directional system.

We conclude with a few interesting patterns in the use of directionals, all of which could be topics for further research.

(1) main stream vs. side stream

In the narratives each of the Ahtna speakers in an automatic way contrasts the main drainage (usually the Copper River) with the tributaries. Jim McKinley's two segments offer a dramatic contrast. As shown in Table 6-3, Jim's first segment lists 71 Ahtna village sites along the Copper River, using the directional 'upstream' 61 times (vs. 'downstream' 13 times and 'upland' 10 times). His second segment mentions 44 place names going up the Klutina River. Here 'upland' is used 42 times whereas 'upstream' is used only twice.

Here are excerpts from each segment with the contrasting 'upstream' vs. 'upland' directionals marked in bold italics:

(A) from Part 1: 'upstream' along the Copper River

Duu yet ***kanii*** k'a xona, yet ***kanii*** xona, T'aghes Tah, T'aghes Tah dae' hwdi'aan see,
/There then, ***the next place upstream***, then ***the next place upstream*** is 'among the cottonwoods', it is named 'among the cottonwoods'.

(B) from Part 2: 'upland' order along the Klutina River

Tl'aticae'e dae' konii de.
/Thus is said 'rear water mouth'.

Yet *kanggat* yełdu' Ts'ekul'uu'i Cae'e dae' konii.
/The *next place upland* of there then is said thus 'one-that-washes-out mouth'.

Jim is consistent with this orientation in these two narrative segments, using 'upstream' along the Copper River but 'upland' for the Klutina River. This is a very strong signal of the impact of the riverine directionals on environmental congnition in Ahtna and other Northern Athabascan languages.

(2) Compounded directionals

The experts add even further precision by compounding two directionals. The most frequently occuring compounds involve 'across' + 'upstream'. 'Across' always comes first: **kanaa 'uniit** 'next place across + distant upstream specific place'; **nanaa 'uniit** 'nearby across + distant upstream specific place'; **nanaa nanuughe** 'nearby across + nearby upstream area.' Also found on occasion are 'across' and other directionals, and again with 'across' coming first: **kanaa 'utsiit** 'next place across + distant downland specific place'; **nanaa 'unggat** 'nearby across + distant upland specific place'. Other compounded directionals, such as **'utggu 'utsiit** 'distantly above + distantly downland', also appear. The ways in which such a powerful extension of the directionals can be used have never been researched through elicitation.

(3) Plain adverb directionals

When the plain form of the directional is used, without prefix and suffix the directional seems to refer to of broad direction or area. Here are a couple of examples in bold face:

(A) Adam Sanford, Part 1

Ts'itaeł Na' łuu ts'e' kudełdiye yehwts'en.
/From 'river that flows straight' to the glaciers is a short distance.
Oh about three miles, two miles, I guess.

Yet xona uyii **naane** skets'enaes.
/Then we would move **across** and into it (canyon) there.

Adam notes that they went from the south side of the Sanford River plain over to the north side, where they access two glacial streams. The word **naane** *makes the tracing of his route clear.*

(B) Adam Sanford, Part 1

Du' yehwts'en ye **nae'** 'stenaes 'utggu daaghe **nae'k'e**
/From there we started out **to the upstream** to above the timberline **on the upstream**.

Adam seems to be referring generally to the direction up the Copper River, as viewed from the Sanford River.

(4) Multiple uses of directionals in a sentence

Experts who know the country well are able to pack a couple of alternate vistas or routes into a single sentence. Here are examples from Jake Tansy and Adam Sanford:

(A) Jake Tansy, Part 2

Satggan łu' **da'aa** Kacaagh **datsiits'en** Kacaagh **datsen 'utsene** hwdaaghe ka'sghideł.
/In the morning **beyond there | on the lowland side** of 'large area' **in the lowland | to the distant lowland** of 'large area' (3-24, Deadman Lake area), we would climb up above timberline.

*Here Jake uses four directionals seemingly in refined ways as he uses the place name **Kacaagh** twice. Perhaps these are phrases conjoined as "or" for alternate potential routes.*

(B) Adam Sanford, Part 1

Ts'itaeł Na' **Ngge' 'ungge na'aa ngge'** 'stedeł dze'.
/In the 'river that flows straight **uplands**' we go **upland | out and over | into the uplands**.

*This sentence has four directionals. Adam seems to be indicating alternative views and areas. The first **Ngge'** is part of the place name **Ts'itaeł Na' Ngge'**.*

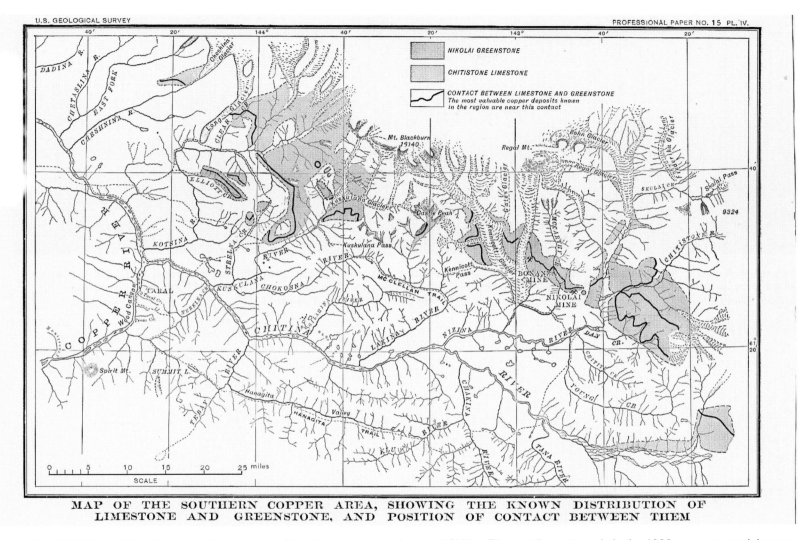

U.S. GEOLOGICAL SURVEY
PROFESSIONAL PAPER NO. 15 PL. IV.

NIKOLAI GREENSTONE

CHITISTONE LIMESTONE

CONTACT BETWEEN LIMESTONE AND GREENSTONE
*The most valuable copper deposits known
in the region are near this contact*

MAP OF THE SOUTHERN COPPER AREA, SHOWING THE KNOWN DISTRIBUTION OF
LIMESTONE AND GREENSTONE, AND POSITION OF CONTACT BETWEEN THEM

Figure 6-3. 1903 Map of the Southern Copper Area. The Ahtna abandoned several Chitina River settlements early in the 1900s as copper mining was developing. The Ahtna presence on the Chitina River is known from place names and trails reported by Andy Brown, Frank Billum, and John Billum Jr. and from some early sources such as Allen 1887, Hayes 1892, Rohn 1899, and Mendenhall and Schrader 1903. As shown on this map of the Chitina and Kotsina drainages by USGS geologists Mendenhall and Schrader (1903, Plate 15), about 20 Ahtna-origin place names had been established. Most, if not all, of the intersecting trail routes around the Chitina River were aboriginal Ahtna trails. For example, prior to 1900 the Taral people used the trail south of Taral up Canyon Creek over to Summit Lake and up the Hanagita Trail (de Laguna n.d., page 1).

Figure 6-4. A view to the east from Slide Mountain, Tits'ikaeni **'sunken boat', looking down the Little Nelchina River** Neltsii Na' **'made in a shape river' with Tazlina Lake** Bendil Bene' **[2-2] 'lake flows lake' in the distance.** This view is along the ancient east-west trail between the Copper River and Knik Arm. The latter two place names were first documented in 1797 by Tarkhanov.

Photo by Suzanne McCarthy.

References

Abercrombie, W. R. 1898. *The Copper River and Adjacent Territory.* Map in 20th Annual Report. Washington: U.S. Government Printing Office.

———. 1900a. A Supplementary Expedition to the Copper River Valley, 1884. In *Compilation of Narratives of the Exploration of Alaska,* pp. 383–411. Washington: U.S. Government Printing Office.

———. 1900b. *Alaska 1899, Copper River Exploring Expedition.* Washington: U.S. Government Printing Office.

Allen, Henry T. 1887. *Report of an Expedition to the Copper, Tanana, and Koyukuk Rivers in the Territory of Alaska in the Year 1885.* Washington: U.S. Government Printing Office.

Austin, Basil. 1968. *The Diary of a Ninety-eighter.* Mount Pleasant, Mich.: J. Cumming.

Billum, John. 1979. *Atna' Yenida'a.* Anchorage: National Bilingual Materials Development Center.

Black, Lydia. 2008. Commentary on Journal of 1796 by Dmitrii Tarkhanov. In *Anooshi Lingit Aani Ka/Russians in Tlingit America: The Battles of Sitka, 1802 and 1804.* Ed. by N.M. Dauenhauer, R. Dauenhauer, and L. Black. Seattle: University of Washington Press. Pp. 67–90.

Bleakley, Geoffrey T. 1998. Historic Properties Associated with the Development of Transportation in Wrangell–St. Elias National Park and Preserve, Alaska, 1885–1955. National Park Service ms. Draft. 21 pp.

Bourke, Joseph 1898–1899. [1899 Sketch map of Copper River]. Bourke Papers, Valdez Museum and Historical Archive Association, Valdez, Alaska.

Brown, C. Michael. 2004. Navigability of Klutina River and Klutina Lake in the Copper River Region. Bureau of Land Management document AA-085087 (1864). 40 pp.

de Laguna, Frederica. 1970. [Sites in Ahtna Territory.] Ms. Typescript. 46 pp.

Hayes, Charles W. 1892. An Expedition Through the Yukon District. *National Geographic.* May 15:117–159.

Irving, William. 1957. An Archaeological Survey of the Susitna Valley. Anthropological Papers of the University of Alaska 6(l):37–52.

Johns, Ruth. 1986. *Ahtna Personal Names.* Copper River Native Association.

Kari, James. 1973–2010. Ahtna fieldnotes summary. 1,300+ pp. in 16 notebooks.

———. 1983. *Ahtna Place Names Lists.* Fairbanks: Copper River Native Association and Alaska

Native Language Center. 105 pp. and 2 wall maps.

———. 1986 (editor). *Tatl'ahwt'aenn Nenn'/The Headwaters People's Country: Narratives of the Upper Ahtna Athabaskans*. Fairbanks: Alaska Native Language Center.

———. 1990. *Ahtna Athabaskan Dictionary*. Fairbanks: Alaska Native Language Center.

———. 1997. *Upper Tanana Place Names Lists and Maps*. Ms. Wrangell-St. Elias National Park. 33 pp. and 10 maps.

———. 1999. Draft Final Report: Native Place Names Mapping in Denali National Park and Preserve. National Park Service.

———. 2004. A Discussion of Three Ethnogeographic Narratives: Nick Kolyaha (of Iliamana), Jim McKinley (of Copper Center), Jake Tansy (of Cantwell). Alaska Native Language Center Working Papers No. 4: 172–79.

———. 2008. *Ahtna Place Names Lists*. 2nd edition revised. Fairbanks: Alaska Native Language Center.

———. to appear. A Case Study in Ahtna Athabascan Geographic Knowledge. In *Landscape in Language: Trans-disciplinary Perspectives*. Mark, D. M., Turk, A. G., Burenhult, N., and Stea, D. (Editors). Amsterdam: John Benjamins Publishing, Culture and Language Use: Studies in Anthropological Linguistics (CLU-SAL) Series.

Kari, James, and James A. Fall. 2003. *Shem Pete's Alaska: The Territory of the Upper Cook Inlet Dena'ina*. Fairbanks: University of Alaska Press. 2nd edition.

Kari, James, and Siri Tuttle. 2005. Copper River Native Places: Report on Culturally Important Places to Alaska Native Tribes of Southcentral Alaska, Bureau of Land Management, Glennallen Field Office.

Marshall, Shelly. 2009. How Tok Got Its Name. *Mukluk News* Vol. 34, No. 23, December 17, 2009.

Mendenhall, Walter C., and Frank C. Schrader. 1903. Mineral Resources of the Mount Wrangell District, Alaska. U.S. Geological Survey Professional Paper 15. Washington: U.S. Government Printing Office.

Moffit, Fred H. 1904. [Map of Matanuska-Susitna rivers, drawn by a local Ahtna-Dena'ina person.] Inside front cover of field notebook 89. U.S. Geological Survey Archives, Menlo Park, CA.

———. 1912. The Headwater Regions of Gulkana and Susitna Rivers. U.S. Geological Survey Bulletin 498. Washington: U.S. Government Printing Office.

Orth, Donald J. 1967. *Dictionary of Alaska Place Names*. Geological Survey Professional Paper 567. U.S. Geological Survey. Washington: Government Printing Office.

Powell, Addison. 1900. Report of Addison M. Powell. In Abercrombie, W. R. 1900. *Alaska 1899, Copper River Exploring Expedition*. Washington: U.S. Government Printing Office, pp. 131–138.

———. 1901. Powell's Map of the Upper Copper River District. Map, Valdez Museum, Valdez, Alaska.

———. 1910. *Camping and Trailing in Alaska*. New York: Wessels & Bissell.

Rand McNally. *Rand McNally Guide to Alaska and Yukon for Tourists, Investors, Homeseekers and Sportsmen*. Chicago: Rand McNally.

Reckord, Holly. 1983. Where Raven Stood: Cultural Resources of the Ahtna Region. Occasional Paper No. 35. Fairbanks: Cooperative Park Studies Unit.

Rohn, Oscar. 1899. A Reconnaissance of the Chitina River and the Skolai Mountains in 1899, Alaska. 21st Annual Report USGS:395-440. Washington: U.S. Government Printing Office.

———. 1900a. Report of Oscar Rohn on Exploration in the Wrangell Mountain district. In Abercrombie, W. R. 1900. *Alaska 1899, Copper River Exploring Expedition*. Washington: U.S. Government Printing Office, pp. 88–94.

———. 1900b. Trails and Routes. In *Compilation of Narratives of the Exploration of Alaska*, pp. 780–84. Washington: U.S. Government Printing Office.

———. 1900c. An Expedition into the Mount Wrangell Region. In *Compilation of Narratives of the Exploration of Alaska*, pp. 790–803. Washington: U.S. Government Printing Office.

Simeone, William E. 2009. Nataełde "Roasted Salmon Place," a Summary History. In *Chasing the Dark, Perspectives on Place, History and Alaska Native Land Claims*. Ed., K. L. Pratt. Anchorage: Bureau of Indian Affairs Alaska Native Claims Settlement Act Office. Pp. 88–95.

Simeone, William E., and James Kari. 2002. Traditional Knowledge and Fishing Practices of the Ahtna of Copper River, Alaska. Alaska Department of Fish and Game, Division of Subsistence Technical Paper No. 270.

———. 2005. The Harvest and Use of Non-salmon Fish Species in the Copper River Basin Office of Subsistence Management Fisheries Resource Monitoring Program. Alaska Department of Fish and Game, Division of Subsistence.

Tansy, Jake. 1982. *Indian Stories, Hwtsaay Hwt'aene Yenida'a, Legends of the Small Timber People*. Transcribed by Louise Tansy Mayo. Anchorage: National Bilingual Materiails Development Center, 89 pp., reprinted in 1997 by Ahtna Heritage Foundation.

Tuttle, Siri G., and James Kari. to appear. *A Seletion of Ahtna Athabascan* Yenida'a *and Cultural Narratives*. Fairbanks: Alaska Native Language Center.

West, Constance F. 1973. An Inventory of Trails and Habitation Sites in the Ahtna Region. Unpublished ms. [Also audio tape collection from that project].